"This book is an incredibly important addition to the global narrative about improving cities. With this thorough academic treatment of data and analysis the conclusion is clear. A shining vision of future transport in cities everywhere is revealed. Not only is it attainable, but it is a thing of beauty."

— *Mikael Colville-Andersen, author of* Copenhagenize:
The Definitive Guide to Bicycle Urbanism

"In this highly interesting and necessary book the authors illustrate the street fight in the presumably perfect cycling city of Copenhagen. Through thorough empirical investigation they argue that Copenhagen can be used as a model for expanding urban cycling not least because of the ideologically political street fights between cars and cycles, identical to other cities around the world. This book, very convincingly, shows the necessary conflictual pathway to thinking about better urban futures."

— *Malene Freudendal-Pedersen, Professor in Urban Planning,*
Aalborg University, Denmark

"In this book Jason Henderson and Natalie Marie Gulsrud have done the world's cities a big favor. They analyze how Copenhagen has made itself bikeable and walkable, not just driveable, and has tempted drivers out of their cars onto bicycles, public transport, or their own two feet. They also explain the politics that produced this result, and show how other cities can follow Copenhagen's wise lead."

— *Donald Shoup, Distinguished Research Professor in the Department of Urban Planning,*
UCLA, USA

STREET FIGHTS IN COPENHAGEN

With 29 percent of all trips made by bicycle, Copenhagen is considered a model of green transport. This book considers the underlying political conditions that enabled cycling to appeal to such a wide range of citizens in Copenhagen and asks how this can be replicated elsewhere.

Despite Copenhagen's global reputation, its success has been a result of a long political struggle and is far from completely secure. Car use in Denmark is increasing, including in Copenhagen's suburbs, and new developments in Copenhagen include more parking for cars. There is a political tension in Copenhagen over the spaces for cycling, the car, and public transit. In considering examples of backlashes and conflicts over street space in Copenhagen, this book argues that the kinds of debates happening in Copenhagen are very similar to the debates regularly occurring in cities throughout the world. This makes Copenhagen more, not less, comparable to many cities around the world, including cities in the United States.

This book will appeal to upper-level undergraduates and graduates in urban geography, city planning, transportation, environmental studies, as well as transportation advocates, urban policy-makers, and anyone concerned about climate change and looking to identify paths forward in their own cities and localities.

Jason Henderson is Professor in the Department of Geography and Environment at San Francisco State University, USA.

Natalie Marie Gulsrud is an Assistant Professor in the Department of Geosciences and Natural Resource Management at the University of Copenhagen, Denmark.

ADVANCES IN URBAN SUSTAINABILITY

Urban Sustainability in Theory and Practice
Circles of Sustainability
Paul James

Sustainability Citizenship in Cities
Theory and Practice
Edited by Ralph Horne, John Fien, Beau B. Beza and Anitra Nelson

Street Fights in Copenhagen
Bicycle and Car Politics in a Green Mobility City
Jason Henderson and Natalie Marie Gulsrud

STREET FIGHTS IN COPENHAGEN

Bicycle and Car Politics in a Green Mobility City

Jason Henderson and Natalie Marie Gulsrud

LONDON AND NEW YORK

First published 2019
by Routledge
2 Park Square, Milton Park, Abingdon, Oxon OX14 4RN

and by Routledge
52 Vanderbilt Avenue, New York, NY 10017

Routledge is an imprint of the Taylor & Francis Group, an informa business

British Library Cataloguing in Publication Data
A catalogue record for this book is available from the British Library

Library of Congress Cataloging-in-Publication Data
Names: Henderson, Jason, 1972- author. | Gulsrud, Natalie Marie.
Title: Street fights in Copenhagen : bicycle and car politics in a green
mobility city / Jason Henderson and Natalie Marie Gulsrud.
Description: Abingdon, Oxon ; New York, NY : Routledge, 2019. |
Series: Advances in urban sustainability | Includes bibliographical references and
index.
Identifiers: LCCN 2019002766 (print) | LCCN 2019015809 (ebook) | ISBN
9780429444135 (eBook) | ISBN 9781138317536 (hbk) | ISBN 9781138334892
(pbk) | ISBN 9780429444135 (ebk)
Subjects: LCSH: Transportation and state--Denmark--Copenhagen. |
Transportation--Environmental aspects--Denmark--Copenhagen. |
Transportation--Denmark--Copenhagen--Planning. |
City traffic--Denmark--Copenhagen. | Cycling--Government
policy--Denmark--Copenhagen. | Transportation, Automotive--Government
policy--Denmark--Copenhagen. | Traffic engineering--Denmark--Copehagen. |
City planning--Environmental aspects--Denmark--Copenhagen.
Classification: LCC HE311.D42 (ebook) | LCC HE311.D42 C6636 2019 (print) |
DDC 388.3/470948913--dc23
LC record available at https://lccn.loc.gov/2019002766

ISBN: 978-1-138-31753-6 (hbk)
ISBN: 978-1-138-33489-2 (pbk)
ISBN: 978-0-429-44413-5 (ebk)

Typeset in Bembo
by Taylor & Francis Books

CONTENTS

ILLUSTRATIONS

Figures

Tables

ACKNOWLEDGEMENTS

Many moving pieces made this book happen. Research commenced in the summer of 2015 with a Development of Research and Creativity (DRC) grant from the Chancellor's Office at California State University. I was able to expand the research considerably when, in 2016, San Francisco State University's Office of the President provided a sabbatical, thus making time for an extended summer and fall semester in Copenhagen. San Francisco State's Office of Research and Sponsored Program's (ORSP) helped to leverage the sabbatical with lodging and travel support and got me back to Copenhagen in summer 2017. Alison Sanders, then Assistant Vice President for ORSP was very helpful, and Yuen Ying Lee, Grant Support Coordinator at ORSP, deserves special acknowledgement for her patience and meticulous oversight.

In Copenhagen, appreciation extends to the Section of Geography at the University of Copenhagen (KU) and the Department of Management, Society and Communication at Copenhagen Business School (CBS). Lars Winther at KU and Eric Guthey and Karl-Heinz Pogner at CBS taught me a lot about the city beyond mobility politics. Eric gets special credit for nudging me to stay longer in Copenhagen.

All of the interviews for this book were conducted while I was on sabbatical in Copenhagen in 2016 and in the summer of 2017. Over fifty planners, advocates and experts in Copenhagen gave their time and offered perspectives either through formal interviews or informal conversation. A few people went above and beyond in sharing their perspectives and insights: Malene Freudendal-Pederson, (now at Aalborg University, previously Roskilde University), Helen Lundgaard at Capital Region Denmark, and Mikael Colville-Andersen and the Copenhagenize crew were all genuinely helpful and welcoming. Two standout politicians, Klaus Bondam and Morten-Kabell, were very gracious with multiple follow-up interviews to help to sort out Copenhagen's mobility puzzle.

Space and time provided by the Institute of Geography, North American Studies, at Heidelberg University in Germany was extremely helpful in the Fall of

2018. My guest professorship at Heidelberg also opened my eyes to the politics of mobility in Germany, and I am humbly indebted to Ulrike Gerhard and Gregg Culver for all their generosity.

From the very beginning of this project, Jerry Davis, Chair of the Department of Geography and Environment at San Francisco State, helped me to patch things together so that I could continue to focus on Copenhagen. For four years Erica Thomas, Theresa Kane, and Alisha Huajardo kept everything moving at the department office. Students in my 'Bicycle Geographies' and 'Global Transportation' courses helped to keep Copenhagen front-and-center between 2015 and 2018. Michael Webster, as always, did stellar cartography.

At Routledge, Rebecca Brennan helped to kickstart the book and Leila Walker shepherded us through the writing. Elizabeth Spicer managed the production. Alison Phillips' copy-editing was meticulous and all the staff at Routledge were gracious and patient.

Of course, special acknowledgement goes to Natalie Marie Gulsrud for her dedicated co-authorship and for helping to steer our conversations and analysis towards hope while avoiding over-romanticizing Copenhagen. Thanks also go to Mary Louis Battalora, my mother, who is a retired English teacher and who insisted on proofreading every chapter. She also introduced me to Europe and for that I am forever grateful. Fellow travellers Andrew Oliphant, Steven Jones, Dave Snyder, Jiro Yamamoto, Will Rostov, and Terry Rolleri patiently listened to my analysis of Copenhagen while cycling the hills of the Bay. Aaron Henderson provided the utilitarian bicycles that took me all over Zealand, the Alps and across California and the West Coast, but also to and from work and for daily transport. Finally, Kate Lefkowitz provided constant enthusiasm and moral support throughout the writing.

Jason Henderson

Acknowledgements for this book are due to my transportation planning community in Seattle, including my dear colleagues at Seattle Children's, Stephanie Innis Frans, Paulo Nunes-Ueno, Barb Culp, Bria Schlottman, Ronna Dansky, Sandy Stutey, Maggie McGehee, Matt Bullen, Corey Holder, and Craig Schneider, who all encouraged me to take the leap and move to Copenhagen. This book is inspired by the team's dedication to making Seattle a more bikeable and just city. The Scan Design Foundation by Jens and Inger Bruun supported my move in 2009 to Copenhagen.

In Copenhagen, acknowledgements are due to my colleagues at the University of Copenhagen (KU) Department of Geosciences and Natural Resource Management, specifically Trine Agervig Carstensen, Anton Stahl Olafsson, Hans Skov-Petersen, Ole H. Caspersen, Lars Winther, and Frank Søndergaard Jensen for their active and helpful decoding of Danish planning culture. Special thanks go to my head of section, Henrik Vejre, for his staunch support of free research time.

Finally, I am indebted to my co-author Jason Henderson for leading and initiating this project, but also for giving me the opportunity to write about my new home. Kasper, Pixie, and Pascal Rasmussen provided unending support and love during this project.

Natalie Marie Gulsrud

INTRODUCTION

Why Copenhagen?

The incredible potential of cycling

Transportation is one of the most vexing and challenging environmental, social, and political problems in the world today. Transport accounts for almost 25 percent of the world's total greenhouse gas emissions, more than double that of the 1970s. Indeed, transport emissions are rising faster than those in any other sector of society, and another doubling is possible by 2050 (Sims *et al.*, 2014). Currently 10 percent of the world's population, largely in North America and Europe, produces 80 percent of the world's transport emissions, with car driving accounting for a large portion of that (ibid.; Replogle and Fulton, 2014). This is deeply inequitable. If cumulative emissions from highway construction, oil extraction, refining, and delivery, and from vehicle manufacturing, are included as part of transport emissions, the emissions from cars and light trucks are substantially greater—and correspondingly alarming (Heede, 2013). In particular, car-oriented nations such as the United States, Germany, and other European neighbors need to address this fairly and swiftly.

Transitioning away from private cars must begin now and cycling—human-pedaled, on two wheels, rubber tires with wire spokes and a diamond frame—can be a pillar of that transition. The Intergovernmental Panel on Climate Change (IPCC)'s *Fifth Assessment Report*, the global standard for what needs to be done, points to the need for strong policies that replace urban car trips with cycling (Sims *et al.*, 2014, p. 624). The need for a transition away from private cars was reinforced by the IPCC's 2018 report entitled *Global Warming of 1.5° C*, which stressed high confidence in an immediate need for rapid and far-reaching transitions in transport, and industries connected with energy, the land, and urban infrastructure. Notably, the report calls for structural changes through urban planning and sustainable consumption lifestyles—such as reconfiguring streets to encourage cycling and creating compact walkable and bicycle-friendly cities. These changes must be undertaken in combination with trends in the adoption of electric cars and renewable energy. Structural changes like compact urbanism and public transport combined with cycling can bring 20–50 percent reductions in urban transport emissions worldwide (IPPC, 2018; see also Replogle and Fulton, 2014).

The world cannot rely solely on technological solutions such as the unproven capacity to electrify one billion cars using unidentified sources of renewable energy (Zehner, 2012). Scaling up renewables for the residential, commercial, and industrial sectors will require massive effort and ingenuity, and renewables are in effect already allotted to these non-transport sectors (Delina, 2016). This points to a need to secure global-scale renewable outlays for non-transport sectors by reducing transport emissions such that private cars (including electric driverless cars) become the last choice, and one that would hardly be needed, for city life and mobility (Wynes and Nicholas, 2017). If current rates of global automobility are pursued, perhaps rising from one billion to two billion cars, we will overshoot the planet's capacity to cope, even if other sectors of emissions are mitigated. We must reduce the need for car travel to enable the deep decarbonization of transport so that global emissions can pass their peak by 2030 (IPCC, 2018).

There exists an incredible potential for shifting many urban trips from the car to the bicycle. Most trips by car in wealthy nations are short and within the spatial range of comfortable cycling. In Europe, half of all car trips are under three miles, and 30 percent are under 1.8 miles (3 kilometers) (World Health Organization, 2014). In the United States, 21 percent of car trips are under one mile, 46 percent are less than three miles, and 60 percent are under five miles (United States Department of Transportation, 2017). These are the kinds of trips that take between 15 and 30 minutes on a bicycle, and there can be a shift towards cycling for many of these trips (Naess, 2006; Tumlin, 2012; Neves and Brand, 2018). In low-income nations, where rapid motorization of a small portion of the elite population is causing traffic chaos and deepening inequity, urban cycling has much to offer (Schipper, 2010; Martin, 2015).

In car-dependent cities, reducing car use and shifting to urban cycling also brings impressive co-benefits including energy security (due to reduced petroleum demand), less air pollution, reduced household costs, fewer fatalities and injuries, less noise-related stress, and improved public health through physical activity (Naess, 2006; Ewing *et al.*, 2008; Sims *et al.*, 2014; Foletta and Henderson, 2016). In the world's developing cities, planning around cycling and public transport can help to avoid the pitfalls of chronic congestion, fatalities and injury, and air pollution experienced in car-based cities (Martin, 2015).

For all cities then, inspiring models for "green" mobility—urban transport centered on reducing car use and increasing cycling, walking, and public transport—should be sought after and understood. In discussing the potential of cycling for climate mitigation, the IPCC points to cities in Denmark, the Netherlands, and Germany as models, and specifically highlights the success of Copenhagen as an iconic bicycle city (Sims *et al.*, 2014).

Indeed, Copenhagen deserves recognition for its bicycle system, and we believe that Copenhagen can provide inspiration and direction as cities around the world look for practical ways to reduce car dependency. Academics and practitioners alike have heralded Copenhagen as a best-practice city for green urbanism and mobility based on the high rate of cycling per citizen in the city and the corresponding level of livability enjoyed by its inhabitants and visitors (Beatley, 2000; Naess, 2006; Pucher and Buehler, 2012; Replogle and Fulton, 2014; Newman and Kenworthy, 2015). However, missing from these assessments is analysis on how becoming a bicycle city is a deeply political pursuit subject to the power struggles and controversies associated with how cars and bicycles should share streets and urban space.

Accordingly, we have chosen to write about Copenhagen as an iconic bicycle city not simply because it is an exceptional or unique outlier, but rather a city with remarkably similar controversies and political issues regarding cars and cycling to those that occur in almost all cities around the world. Invoking a politics of mobility framework we consider how political power, expressed through political ideology and operationalized by political parties and organizations, shapes urban transport policy in Copenhagen (on the politics of mobility framework see Cresswell, 2010; Henderson, 2013; Sheller, 2014). By political ideology we mean a set of beliefs and principles about politics and government that include the scope of government, how decisions should be made, and what values government should pursue. In that vein, we place particular emphasis on Copenhagen's ideological differences in debates over where cars and bicycles belong on specific city streets, how many cars and bicycles should be accommodated in the city as a whole, as well as where cars and bicycles should go as they circulate around the city and seek a place to park. These "street fights" are ideologically charged political struggles over street space and other urban spaces that directly or indirectly impact mobility. Flashpoints include how street space is physically allocated, debates over pricing and metering the car, the spaces allocated to parking for cars, and broader debates about how bicycles, cars, pedestrians, and public transit should circulate in the city, by how much, and where exactly and in what numbers (Henderson, 2013).

A central premise of the book is that the politics of mobility travels and Copenhagen is having similar debates about streets and urban spaces to those being held in cities around the world. Ideology also travels. While variegated and shaped by conditions in the locality, we argue that three broad ideologies of mobility— "Left/Progressive," "Neoliberal," and "Right/Conservative"—are present in Copenhagen's politics of mobility and that this is strikingly similar, albeit nuanced and variegated, to many North American and European peer cities. Simultaneously, and also variegated, these ideologies of mobility play out in many cities across the globe.

We will unpack these three ideologies throughout the book, but, briefly, what we call a Left/Progressive politics of mobility challenges the car through the promotion of strong government intervention such that green modes of transportation like cycling and public transit are prioritized. In Copenhagen, a group of political parties—Enhedslisten—de Rød-Grønne (Unity List—the Red-Green Alliance), the Socialistisk Folkeparti (Socialist Peoples' Party), and the new upstart Alternativet (The Alternative)—reflect this ideological position and see the bicycle as core to the urban future. By left we refer to a distinctive articulation challenging the structure of capitalism, whether involving private property or corporate regulation, and we invoke Progressive to stress the value that government can and should do good things for people, and that public solutions should come before private. In that vein, Copenhagen's Left/Progressive politics of mobility articulates government policies that makes it harder to drive, and thus discourages private car ownership. It also prioritizes public spaces and public and collective solutions to urban problems. The challenge to the private car is a key characteristic that we will highlight.

Copenhagen also has a Neoliberal politics of mobility centered on market-based and technocratic approaches. In this case, and probably to many readers' surprise, the traditionally left-leaning Socialdemokratiet (Social Democratic party), allied with Venstre (the traditional business-oriented Liberal party) promotes a politics of mobility in which transport policy should enhance market-oriented economic growth and especially private profit. Neoliberals in Copenhagen simultaneously celebrate the bicycle as an attractive green marketing instrument, but, distinctive from the Left/Progressives, the Neoliberals also articulate policies accommodating more cars in Copenhagen. With limited street space this brings many contradictions and ambivalence into Copenhagen's politics of mobility because bicycling is compromised for more car space.

Rounding out the politics of mobility in Copenhagen are the Right/Conservatives, such as Det Konservative Folkeparti (Conservative People's Party) and the right-wing populist Danske Folkeparti (Danish People's Party). The Right/Conservatives fuse the automobile with politically conservative notions of individualism and personal responsibility with right-leaning populist discourses that essentialize the car as natural, inevitable, and especially necessary for families with children. The role of government, as with the Left/Progressives, is active. Yet the policies are meant to sustain and expand the car system, through the provision of infrastructure and through reducing the cost of driving. For the Right/Conservatives, government should guarantee and accommodate automobility and not seek to discourage it or make it more expensive. Cycling certainly has an appeal in its frugality, individual propulsion, and independent streak, but cycling spaces should not be made available at the expense of car spaces. Right/Conservative political power is largely outside the municipality of Copenhagen but is embedded in Danish national politics. Because the national government exerts power over the city in terms of taxes, transport finance and other important policies such as parking, Right/Conservative politics permeates the city, and the Right/Conservatives frequently ally themselves with the Neoliberals on policies that preserve and expand car access at the expense of cycling and green mobility.

We acknowledge that these categorizations are not always a clean fit and that there can sometimes be grey areas. For example, a small party called Det Radikale

Venstre (Social Liberals, who historically comprise left-leaning urban intellectuals and small land owners) also tacks Neoliberal at times but has a long history of supporting cycling and aligning with Left/Progressives to limit the car. Similarly, the Liberal Alliance, also a small party, is unabashedly Neoliberal but invokes Right/Conservative discourses promoting the car, such as the car being essential for individual freedom and for families with children. The Social Democrats, Denmark's largest party, have leaned Left/Progressive on mobility at key historical moments, but in recent years have tilted towards a Neoliberal and Right/Conservative stance on car policies, preferring at times to accommodate more cars rather than implement parking and congestion pricing policies that would favor cycling (and public transit).

We also acknowledge that Denmark does not have a politically powerful automobile manufacturing sector that in the United States, Germany, and the People's Republic of China, had a historically outsized role in shaping transport policy (Carstensen and Ebert, 2012; Koglin, 2015a, 2015b). Denmark's lack of a national automotive industry made cars easier to tax, and its high car tax reduced car ownership compared to peer countries (Boge, 2006). While the political economy of the car industry is an important factor, it is equally relevant that Copenhagen shows us what a modern city might be like without the political influence of a national car industry (although there are automobile dealers, suppliers, and road builders in Denmark). Copenhagen gives us a refreshing lens into what the politics of possibilities for cycling and green mobility could be in China, Germany, India, Brazil, the United States, and many other nations if their powerful automobile manufacturing interests were politically less influential, their pro-car policies checked, car subsidies were reduced, and private car companies were otherwise unable to disproportionately exert a dominating power over mobility politics. This is not to say that Copenhagen lacks political influence by pro-car interests—this does happen—but that this political influence has been successfully challenged and there are many openings for Copenhagen to strive for more cycling, public transit, and a compact city centered on green mobility.

By way of considering Copenhagen's Left/Progressive, Neoliberal, and Right/Conservative politics of mobility, we seek to highlight the similarities rather than Copenhagen's uniqueness, because we believe that Copenhagen's cycling system can and must be replicated worldwide if we are to meaningfully and equitably address global warming and the corollary environmental, social and economic disorder of private cars. But why Copenhagen? And why should this city's hard-fought for cycling system be replicated worldwide?

Why Copenhagen?

With 29 percent of all trips made by bicycle, Copenhagen is a model of green mobility. Copenhagen appears in global scholarship on sustainable mobility, dozens of foreign delegations flock to Copenhagen for study tours every year, and cycling is used in promoting local economic development and tourism. Crisscrossed by networks of cycle tracks, which are wide, fully separated pathways that provide safe, welcoming travel spaces for bicyclists of all skill levels and ages, Copenhagen's rate of cycling rises to 62 percent of all weekday trips to places of work and education within the city, and bicycles outnumber cars in the urban core (City of Copenhagen, 2017). With the exception of the peer cities Amsterdam and Utrecht in the Netherlands, no other large city in the world approaches Copenhagen, and Copenhagen's reputation as an iconic bicycle city is not an overstatement.

The flip side of high rates of cycling in Copenhagen is relatively low car ownership and low rates of driving. The Municipality of Copenhagen reports under 200 cars per 1,000 persons, a common metric for comparing car ownership rates between cities and countries (City of Copenhagen, 2017). This places Copenhagen at the low end of car ownership compared to peer European cities (averaging roughly 500 per 1,000 persons), and far below cities in the United States, which average over 700 per 1,000 persons (Jones, 2008; Sperling and Gordon, 2009; Newman and Kenworthy, 2015; Eurostat, 2018; OPEC, 2018). Copenhagen's rate of car ownership in 2016 was less than the car ownership rate of the United States during the 1930s (Jones, 2008).

Moreover, 70 percent of Copenhagen households are car free. Household car ownership in Copenhagen hovers at around 30 percent, a particularly low rate for such a wealthy city, while trips made by cars make up 34 percent of all journeys in the city. It is important to point out that most of these car trips are suburban in either origin or destination. When removing suburban car trips from the calculation, and looking only at journeys to work and places of education within the city, cars account for only 9 percent of all trips (City of Copenhagen, 2017). Just as Copenhagen's rates of cycling are among the highest in the world, these low rates of car ownership and car use are impressive, inspirational, and important.

High cycling rates and low car ownership and use have helped Copenhagen to achieve some of the lowest per capita car emissions (0.575 metric tons of CO_2 equivalent emissions per capita in 2015) among peer cities around the world (City of Copenhagen, 2016). Take, for comparison, San Francisco, an important green mobility bellwether in the United States, and with a similar population density (18,500 persons per square mile) to Copenhagen, where per capita emissions from cars were 2.1 metric tons of CO_2 equivalent (San Francisco Department of the Environment, 2017). San Francisco, despite relatively high walking and transit mode shares for an American city, still has high rates of driving and over four times the rate of per capita emissions compared to Copenhagen.

What leads to high rates of cycling and low rates of car use in Copenhagen? Notable transport scholars such as Pucher and Buehler (2012), who highlighted Copenhagen's high rates of cycling in their globally comprehensive examination of urban cycling, emphasize car restraint, or the discouraging of car use through taxes and fees on cars, and decreasing parking and roadway capacity for cars. They are among many scholars, consultants, and advocates who conclude that robust car restraint policies are necessary in order to increase cycling (and transit) mode shares, and make cities around the world greener, livable, and more equitable. High population density by itself, which is considered a key precursor to green mobility, is not the panacea, given that many high-density cities around the world are also crammed with cars and have low cycling rates (ibid.; Martin, 2015). At any density, car restraint and adequate cycling spaces are what is critical, and Copenhagen shows that limited car space is fundamental. Yet while Copenhagen is a globally recognized bicycle city (as well as a green leader in sectors such as wind energy and district heating), less is known about car restraint in Copenhagen, and especially the political tensions between providing bicycle space and reducing spaces occupied by cars. In this book we focus on that omission.

Cycling levels and other sustainable mobility measures are sturdily anchored in Copenhagen, but it has not always been this way, and presently they are threatened (a topic that we will address in this book). Growth in cycling appears to be plateauing due to spatial limitations on the cycle tracks and on neighborhood streets, ironically because Copenhagen, after decades of political work and compromise, put down excellent infrastructure and marketed cycling in a positive image. Now Copenhagen struggles with even harder politics—continuing and expanding car restraint. The struggle is pivotal to Copenhagen's future role as an iconic green mobility capital because new residents are moving to the city with new cars, demanding more street space, while commuters and visitors from Copenhagen's suburbs and beyond are increasingly automobile-oriented (Naess, 2006; Gössling, 2013).

To borrow a line from Shakespeare's *Hamlet*, "Something is rotten in the state of Denmark" (2009, Act 1, Scene IV, p. 49). Flashpoints in Copenhagen's street fights include a proposed congestion pricing scheme intended to reduce the numbers of cars in the city while meeting its climate mitigation goals. The city's congestion pricing scheme, called a "toll ring" and now in limbo (see Chapter 5 in this volume), formed the basis of Copenhagen's Left/Progressive plans for reducing car volumes and greenhouse gas emissions from transportation. Without the toll ring, Copenhagen has been backpedaling on its climate goals while suburban cars have swarmed through the city.

Other flashpoints include conflicts over car parking versus bicycle space in the city's urban center, which includes the medieval urban core (Indre By) of Copenhagen, the nearby harbor redevelopment areas, and the Brokvarterene (Bridge Quarters), the semi-circle of high-density apartment blocks that surrounds Indre By (See Figure 0.1). Parking for cars takes up precious urban space that bicycles also claim in Indre By and the Brokvarterene, while new upscale developments in the harbor and in one of Copenhagen's newest neighborhoods, Ørestad, include parking, which generates more car trips in Copenhagen. With parking for more cars, Copenhagen's Left/Progressive politics of mobility may also shift rightward as more Neoliberal and Right/Conservative discourses about cars come to the forefront of city politics.

Parking for cars runs up against the need to expand and improve the bicycle system, which daily experiences capacity limits and a possible plateauing of Copenhagen's cycling rates. The city may not be able to reach its self-imposed

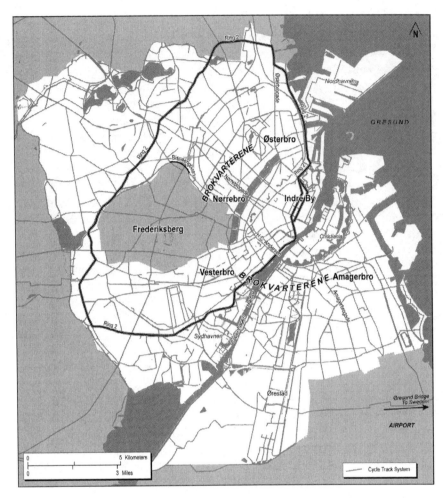

FIGURE 0.1 Overview map of Copenhagen showing cycle track system.
Source: Map by Michael Webster.

target of 50 percent of all work and education trips, including commuters into the city, being taken by bicycle. A big part of that is lack of comfortable, extra-wide cycle tracks and other facilities that would necessarily intrude into existing car space.

In addition to congestion pricing and parking, other flashpoints in the politics of mobility have also emerged, including the proposed construction of a tolled harbor tunnel and a ring road on the eastern flank of the urban core that connects the wealthy (and more right-leaning) northern suburbs of Copenhagen with new upscale developments on Amager, while also providing quicker access by car to Copenhagen's international airport. This 12-kilometer (7.5 mile) bypass tunnel would effectively complete an inner ring motorway around the city, an idea that has been debated for decades. Today the harbor tunnel is part of a broader constellation of local debates over the car and the city, including proposals to close off more of the city center from car traffic and to remove a vestigial segment of elevated urban motorway known as the Bispeengbuen. Related flashpoints include upgrading the venerable S-tog (S-train) regional railway, as well as how to protect the regional public bus system. The anticipated Metro City Ring line, set to commence operations in 2019, might impact cycling, but it will also lead to a restructuring of some city bus routes.

Both the harbor tunnel and Metro City Ring line cut into the heart of the politics of bicycles versus car space. The tunnel, according to Neoliberal and Right/Conservative proponents, might remove cars (and trucks) from parts of the urban core, making it better for cycling. However, Left/Progressive opponents counter that it will simply invite more car traffic, while diverting transportation investment away from greener projects. With the Metro, there are fears that once trains are circling underneath the urban core, cycling rates might decline because of mode shift to the Metro. Added to that, there is concern that once the construction concludes, cars will fill spaces currently restricted to them. Furthermore, there is a crosscurrent debate over restructuring city bus routes when the Metro opens. The interplay of public transportation with the street fights between cycling and the car is important to this analysis.

Throughout the book we will also consider increased car ownership and car use beyond the municipality of Copenhagen. Conventional wisdom holds that Denmark's famously high car tax, at 150 percent of the value of a car in 2018, kept car ownership rates historically low in Denmark, including in Copenhagen's suburbs. That has unraveled in the past decade, and lowering the tax is another political flashpoint in Danish politics, especially for Neoliberals and Right/Conservatives seeking to make it cheaper to own a car. Some political parties such as the Liberal Alliance advocate reducing the car tax explicitly to encourage greater car ownership in Denmark. With greater car ownership outside the municipality of Copenhagen, suburban motorists and motoring tourists exert political pressure for car access into the city, aggravating the debates about congestion pricing, parking, the harbor tunnel, and public transportation—and ultimately, bicycle space.

In sum, these flashpoints show us that Copenhagen is literally at a major crossroad in terms of its leadership as a green mobility capital. Cycling rates, although exceptionally high by world standards, could go higher if the space of the city is steered toward that. But car use and demand for car space could prevail, and the bicycle mode share could very well decline. Copenhagen might then continue to drift away from its climate goals, and perhaps have to relinquish its top ranking as a green mobility capital.

The politics of cycling and the car is not unique to Copenhagen, and is in fact prevalent in cities worldwide. We argue that Copenhagen is more, not less, normal, when compared to other cities. Take, for example, San Francisco, which is one of the green mobility leaders among peer cities in the United States, and which has relatively lower car ownership and car mode shares. In Copenhagen and San Francisco there are similar political and ideological debates over cycling, car parking, and using pricing to manage cars (see Henderson, 2013). Similarly, in Vienna (Austria), and Toronto (Canada), there are political alignments from left to right regarding car parking and pricing (Buehler et al., 2017; Walks, 2014). To list just a few representative cases, cities from Beijing to Mexico City are grappling with these debates albeit in variegated and nuanced ways, as are Barcelona, Berlin, Frankfurt, Hamburg, London, and Paris in Europe, and Los Angeles, New York, Portland, and Seattle in the United States. Therefore, throughout the book we will offer snapshots of comparison between Copenhagen and other cities in North America, Europe, Asia, Africa, and Latin America.

The plan of the book

The book will unfold in the following sequence of chapters. In Chapter 1 we will ask, by way of examining Copenhagen, what exactly is a bicycle city? We will dig deeper into what it means to be an iconic bicycle city, and review in detail the data and metrics pointing to Copenhagen's impressive successes. We suggest that, just as automobile and transit systems have global standards and best practices for infrastructures and urban form, so too must cycling. Data and metrics will also offer comparisons to other cities around the world. Chapter 2 will review the history of Copenhagen's rise as a bicycle city and take Copenhagen's history of cycling and the car up to the present, with the "bicycle renaissance" starting around 2005. In the historical analysis we consider how the politics and ideology of mobility can be factored in, and especially the relationship in Denmark between cycling and social democracy. Chapter 3, titled "Something is rotten in the state of Denmark," considers some of the problems encountered by Copenhagen's cycling system, and points to the rise of car ownership and debates that arise about urban space in Copenhagen.

Next, in Chapter 4 we detail the political ideologies of the Left/Progressives, the Neoliberals, and the Right/Conservatives which encompass Copenhagen's politics of mobility. Chapters 5, 6 and 7, will detail the flashpoints of the toll ring, parking standards and pricing, and mega-projects such as the harbor tunnel and the Metro City Ring, as well as other noteworthy proposals to restructure the public bus system, remove the elevated Bispeengbuen motorway viaduct, and expand the car-free zone in the center of Copenhagen. Our conclusion in Chapter 8 will pull it all together and consider how Copenhagen remains a hopeful city. It will also ask what Copenhagen might do to overcome these contentious street fights and maintain its stature as an iconic bicycle city and green mobility capital.

Copenhagen can be a beacon of hope for a more positive attitude towards the future possibilities for our cities and our climate. We are confident that Copenhagen's high rates of cycling can be achieved anywhere in the world and that the path towards bicycle cities can be cleared of obstacles if we understand what those obstacles are, and that they can be overcome. To reiterate our emphasis, the political debates over bicycle and car space in Copenhagen are not unique. Copenhagen is facing a similar kind of crossroads found in cities around the world. There is much that experts on climate policy, transportation scholars, urban planners, green mobility advocates, and urban policy-makers can learn from Copenhagen's street fights. By considering how Copenhagen's politics of mobility is remarkably similar to the politics of mobility found in other countries, and especially in the car-saturated United States, many parts of Europe, and many modernizing Asian, African, and Latin American cities, we hope that new ways of thinking enthusiastically about the future can be found.

References

Beatley, T. 2000. *Green Urbanism: Learning from European Cities*. Washington, DC, Island Press.

Boge, K. 2006. *Votes Count but the Number of Seats Decides: A Comparative Historical Case Study of 20th Century Danish, Swedish and Norwegian Road Policy*. Oslo, Norwegian School of Management, Department of Innovation and Economic Organization. Ph.D.: 465.

Buehler, R., Pucher, J., and Altshuler, A. 2017. "Vienna's Path to Sustainable Transport." *International Journal of Sustainable Transportation*, 11: 257–271.

Carstensen, T. A. and Ebert, A. K. 2012. "Cycling Cultures in Northern Europe: From 'Golden Age' to 'Renaissance'". In J. Parkin, ed., *Transport and Sustainability*, vol. 1: *Cycling and Sustainability*. Bradford, Emerald Group: 23–58.

City of Copenhagen. 2016. *CPH 2025 Climate Plan: Roadmap 2017–2020*. Copenhagen, Technical and Environmental Administration: 44.

City of Copenhagen. 2017. *Copenhagen City of Cyclists: The Bicycle Count 2016*. Copenhagen, Technical and Environmental Administration: 24.

Cresswell, T. 2010. "Towards a Politics of Mobility." *Environment and Planning D: Society and Space*, 28(1): 17–31.

Delina, L. 2016. *Strategies for Rapid Climate Mitigation: Wartime Mobilization as a Model for Action*. New York, Routledge.

Eurostat. 2018. *Passenger Cars in the EU*. Available at https://ec.europa.eu/eurostat/statistics-explained/index.php?title=Passenger_cars_in_the_EU (accessed 26 February 2019).

Ewing, R., Bartholomew, K., Winkelman, S., Walters, J., and Chen, D. 2008. *Growing Cooler: The Evidence on Urban Development and Climate Change*. Washington, DC, Urban Land Institute.

Foletta, N. and Henderson, J. 2016. *Low Car(Bon) Communities: Inspiring Car-Free and Car-Lite Urban Futures*. New York, Routledge.

Gössling, S. 2013. "Urban Transport Transitions: Copenhagen, City of Cyclists." *Journal of Transport Geography*, 33: 196–206.

Heede, R. 2013. "Tracing Anthropogenic Carbon Dioxide and Methane Emissions to Fossil Fuel and Cement Producers, 1854–2010." *Climatic Change*, 122(1–2): 229–241.

Henderson, J. 2013. *Street Fight: The Politics of Mobility in San Francisco*. Amherst, University of Massachusetts Press.

Intergovernmental Panel on Climate Change (IPCC). 2018. *Global Warming of 1.5 °C*. Geneva, United Nations.

Jones, D. W. 2008. *Mass Motorization and Mass Transit: An American History and Policy Analysis*. Bloomington, Indiana University Press.

Jones, C. and Kammen, D. M. 2014. "Spatial Distribution of U.S. Household Carbon Footprints Reveals Suburbanization Undermines Greenhouse Gas Benefits of Urban Population Density." *Environmental Science and Technology*, 48(2): 895–902.

Koglin, T. 2015a. "Vélomobility and the Politics of Transport Planning." *GeoJournal*, 80(4): 569–586.

Koglin, T. 2015b. "Organisation Does Matter: Planning for Cycling in Stockholm and Copenhagen." *Transport Policy*, 39(April): 55–62.

Martin, G. 2015. *Global Automobility and Social Ecological Sustainability of the Urban Political Economy and Ecology of Automobility: Driving Cities, Driving Inequality, Driving Politics.* Ed. Alan Walks. New York, Routledge: 23–37.

Naess, P. 2006. *Urban Structure Matters: Residential Location, Car Dependence, and Travel Behavior.* London, Routledge.

Neves, A. and Brand, C. 2018. "Assessing the Potential for Carbon Emissions Savings from Replacing Short Car Trips with Walking and Cycling Using a Mixed GPS-Travel Diary Approach." *Transportation Research Part A: Policy and Practice.* Available at www.sciencedirect.com/science/article/pii/S0965856417316117 (accessed 26 February 2019).

Newman, P. and Kenworthy, J. 2015. *The End of Automobile Dependence: How Cities are Moving Beyond Car Based Planning.* Washington, DC, Island Press.

Organization of the Petroleum Exporting Countries (OPEC). 2017. *World Oil Outlook 2040.* Vienna, OPEC: 364.

Pucher, J. and Buehler, R. 2012. *City Cycling.* Cambridge, MA, MIT Press.

Replogle, M. and Fulton, L. M. 2014. *A Global High Shift Scenario: Impacts and Potential for more Public Transport, Walking, and Cycling with Lower Car Use.* New York and Davis, CA, Institute for Transportation and Development Policy and University of California, Davis: 35.

San Francisco Department of the Environment. 2017. *2015 San Francisco Greenhouse Gas Emissions Inventory at a Glance.* San Francisco, CA, San Francisco Department of the Environment. Schipper, L. 2010. "Car Crazy: The Perils of Asia's Hyper-Motorization." *Journal of the East Asia Foundation,* 4(4). Available at www.globalasia.org/v4no4/cover/car-crazy-the-perils-of-asias-hyper-motorization_lee-schipper (accessed 26 February 2019).

Shakespeare, W. 2009. *Hamlet,* Auckland, Floating Press.

Sheller, M. 2014. "The New Mobilities Paradigm for a Live Sociology." *Current Sociology Review* 62 (6): 1–23.

Sims, R., Schaeffer, R., Creutzig, F., Cruz-Núñez, X., D'Agosto, M., Dimitriu, D., Figueroa Meza, M.J., Fulton, L., Kobayashi, S., Lah, O., McKinnon, A., Newman, P., Ouyang, M., Schauer, J.J., Sperling, D., and Tiwari, G. 2014. "Transport." In O. Edenhofer, R. Pichs-Madruga, Y. Sokona, E. Farahani, S. Kadner, K. Seyboth, A. Adler, I. Baum, S. Brunner, P. Eickemeier, B. Kriemann, J. Savolainen, S. Schlömer, C. von Stechow, T. Zwickel and J.C. Minx, eds., *Climate Change 2014: Mitigation of Climate Change. Contribution of Working Group III to the Fifth Assessment Report of the Intergovernmental Panel on Climate Change.* Cambridge and New York, Cambridge University Press.

Sperling, D. and Gordon, D. 2009. *Two Billion Cars: Driving towards Sustainability.* Oxford, Oxford University Press.

Tumlin, J. 2012. *Sustainable Transportation Planning: Tools for Creating Vibrant, Healthy, and Resilient Communities.* Hoboken, NJ, John Wiley and Sons.

United States Department of Transportation. 2017. *2017 National Household Travel Survey.* Washington, DC, Federal Highway Administration. Available at https://nhts.ornl.gov/ (accessed 3 January 2019).

Walks, A. 2014. "Stopping the 'War on the Car': Neoliberalism, Fordism, and the Politics of Automobility in Toronto." *Mobilities,* 10(3): 402–422.

World Health Organization (WHO). 2014. *Unlocking New Opportunities: Jobs in Green and Healthy Transport.* Copenhagen, World Health Organization.

Wynes, S. and Nicholas, Kimberly A. 2017. "The Climate Mitigation Gap: Education and Government Recommendations Miss the Most Effective Individual Actions. "*Environmental Research Letters,* 12. Available at http://iopscience.iop.org/article/10.1088/1748-9326/aa7541/meta (accessed 28 January 2019).

Zehner, O. 2012. *Green Illusions: The Dirty Secret of Clean Energy and the Future of Environmentalism.* Lincoln, University of Nebraska Press.

1

COPENHAGEN

Bicycle city

Observing cycling in Copenhagen

Copenhagen's Dronning Louises Bro (Queen Louise Bridge), which straddles the lakes between Nørrebro and Indre By, has one of the highest concentrations of bicycle traffic in the world. On an average weekday 48,500 cyclists crossed in 2016 (City of Copenhagen, 2017a). To observe the daily ballet of crisscrossing waves of cyclists it is best to visit the Queen Louise Bridge during the morning commute between 8 a.m. and 9 a.m. when children being shuttled to school mix with thousands of commuting workers. During the political season the morning cyclists are greeted by sign-waving pundits handing out campaign fliers.

From Nørrebro, Nørrebrogade is the main trunk line for cyclists, with an average of over 42,000 bicycles a day, and just 7,500 cars (Colville-Andersen, 2015). In the early 2000s cycling rates increased to such an extent that private cars were restricted for several blocks along Nørrebrogade and were forced to turn off the main street. The restriction of cars facilitated the reallocation of car space, shifting it into cycling space, and the cycle tracks were widened so that now a pair of socializing cyclists can be passed easily by a third. This "passability" scheme, described later in this volume, is called "Plus-Net" in local planning parlance. More space was also allocated for Copenhagen's busiest public bus, the 5C, which carries 61,000 passengers daily and was recently upgraded to offer a faster public transit priority service.

Three-quarters of a kilometer south of Queen Louise Bridge, Denmark's busiest car street, Hans Christian Andersens Boulevard (from here on H. C. Andersens), carries 55,000 motor vehicles per day. The eight lanes of cars and trucks resemble the arterial roadways found in cities in Germany or the United States. The difference is that H. C. Andersens has smooth, wide cycle tracks flanking the motor traffic, and 40,000 cyclists cross the junction with the lakes each day. Nearby a

FIGURE 1.1 Image of the morning commute in Copenhagen. Queen Louise Bridge, 2015.
Source: Photograph by J. Henderson.

digital traffic information sign tells eastbound motorists that cyclists will probably reach the Langebro, which is 2.1 kilometers on the other side of central Copenhagen, a few minutes sooner than the cars. Cars move slowly in Copenhagen.

On the other side of the city center at the Knippelsbro (see Figure 1.2), 40,000 bikes stream over the harbor each day from Torvegade which connects Christianshavn and Amager. Two more bridges to the south, Langebro and Bryggebroen, also connecting Amager to the city center, each carry 40,400 and 23,800 cyclists, respectively, while to the north of Knippelsbro, the newer Inner Harbor bridge has begun to relieve congestion on the Knippelsbro and currently carries some 17,500 cyclists per day (City of Copenhagen, 2017a).

Each weekday in Copenhagen roughly 260,000 cyclists cross between the Brokvarterene (Bridge Quarters) to Indre By (City of Copenhagen, 2017b). These high-density residential neighborhoods—Indre By and Christiania in the center, surrounded in a semi-circle by Nørrebro, Østerbro, and Vesterbro on the western side of the lakes, and Amagerbro east of Copenhagen harbor (see Figure 1.2)— have the urban design, density, and human scale that throughout the twentieth century encouraged Copenhageners to take to their bicycles. It was here that mass cycling took off in Copenhagen during the early 1900s, and where the built environment of Copenhagen was clearly not designed to favour the car. Flat, compact, and politically disposed to cycling and car restraint, the Indre By and

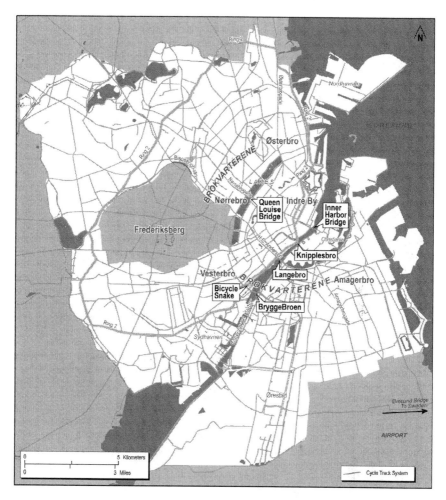

FIGURE 1.2 Map of Copenhagen's cycle track system and bicycle bridges over Copenhagen harbor.

Source: Map by Michael Webster.

Brokvarterene were—and remain today—an optimal cycling environment and model of green mobility (Emanuel, 2016).

Within Indre By's well-preserved medieval and Renaissance layout the bicycle fits easily onto the narrow side streets. In the surrounding semi-circle of the Brokvarterene, built between the 1870s and the 1920s, one can easily and safely cycle in under 20 minutes to the center of Copenhagen, to hundreds of small businesses and supermarkets, and to schools, parks, and nature reserves. Excellent rail connections on the S-train, the regional commuter rail system, expand the cyclists' spatial range to the entire Copenhagen metropolitan region. If Copenhagen strives to be crowned a green mobility capital, the area between the harbor and the Ring 2 (circling the urban core west of the harbor), with a sizeable portion of northwestern Amager, is the crown jewel.

The car is arguably not part of this "tout ensemble" (assemblage of parts and details). In the Brokvarterene and Indre By apartments do not come with off-street parking for cars (although parts of some courtyards have been converted for parking). However, with livable density and intense mixing of uses, there is little room for mass car storage here. The dearth of off-street parking means few curb cuts and driveways splaying across cycle tracks. Most of the cycle tracks line up adjacent to the sidewalk, and streets such as Nørrebrogade and Vesterbrogade have no curbside car parking.

All of this makes it easier to construct and widen cycle tracks and to prioritize cycling. It also helps to explain low car ownership rates. In Nørrebro there are 115 cars per 1,000 persons, and in Vesterbro 140 cars per 1,000 persons (City of Copenhagen, 2016). Overall, approximately 70 percent of households in Copenhagen are car free. As of 2015, each weekday more bicycles (260,000) than cars (250,000) pass through the Brokvarterene to Indre By (ibid., 2017b). In absolute numbers, cars are declining and cycling increasing in Copenhagen's urban core, and since 2007, at the start of Copenhagen's current cycling renaissance, cycling has increased by 51 percent in the core (ibid., 2017a).

The mainline cycle tracks of the Brokvarterene and Indre By are also good places to observe the cycling style and behavior of Copenhageners. In Copenhagen, it is said that cyclists are less reckless and more orderly on weekday mornings, when they are more narrowly focused on their surroundings because the cycle tracks are so dense with other cyclists (cyclists are more spread out and cycling behavior can be a little looser during the evening commute). Scofflaw maneuvers such as hopping out of the cycle track in order to pass other cyclists are rare. Instead, and to keep things efficient, social solidarity is observable, and there is no cutting into the queues that, at main junctions like the Queen Louise Bridge at the lakes could extend 25 or 50 cyclists back by 8.15 a.m. Cyclists queue up patiently even if they might have to wait for two traffic light cycles before crossing. There are few bells ringing aggressively and impatiently (although gentle bell-ringing is regarded as a courteous gesture when passing), and there is little arguing.

From the Queen Louise Bridge, some cyclists making a "Copenhagen left," or a two-stage box turn, make a fishhook turning maneuver, sweeping slightly against the perpendicular traffic flow to point north along the east embankment of the lake. The "Copenhagen style" is on display here and ordinary, commonplace bicycles are juxtaposed with sophisticated fashionable cyclists. Many of the bicycles are rickety and appear poorly maintained, and are interspersed with new "Long John" cargo bikes ferrying children, hip vintage bikes, and a smattering of road and mountain bikes. The vast majority are nondescript black bikes with wide fenders and often a basket in front or back.

There is little heavy breathing or sweating. A few long-distance cyclists, usually commuting from the suburbs, wear Spandex and clothing designed specifically for cycling. Most cyclists, however, wear their work clothes: dress pants, dresses and skirts, trench coats in the rain, black shoes, and basic work clothing. Few cyclists wear helmets although helmet use might be on the rise. Many cyclists might have one hand steering the bicycle and the second hand holding coffee or a smartphone. Children are sometimes nestled on the front bars, but in Copenhagen you do not see too many people piled onto one bicycle as you might see in Amsterdam.

Stand and watch longer, and you will observe subtle behavior and movements, such as nonchalant cyclists cocking their heads slightly and glancing back when passing, or subtly glancing back and down to the right when summoning a fellow cyclist to ride side by side. Hand signaling is obligatory but is not done vigorously or with overly gesticulating hand motions. Instead arms are extended casually, not briskly, with a relaxed finger pointing downward in the direction of the turn.

Observing cycling in Copenhagen includes listening. Motorized traffic noise can be intermittent, even during rush hour. At regular intervals after a small set of motor vehicles clears it can be especially quiet on Queen Louise Bridge and Nørrebrogade as dozens of cyclists continue stealthily for minutes at a time. One hears only spinning chains and gears, and mildly squeaking brakes, against a backdrop of low-whispered conversations between sociable cyclists riding side by side. Copenhagen is particularly quiet after the evening rush hour, when most cars have driven back to the suburbs.

Watch more closely and the rules start to become apparent. Cycling is not allowed on pedestrian crosswalks, but a pair of cyclists might walk their bikes on the crosswalk while engaged in conversation or to use their phones. Because there is so much cycling cross-traffic at intersections, right turns on red are restricted, and neither bicycles nor cars should do it, although many cyclists would like the right on red legalized, and there is a low-level of "gentle rule breaking" (Larsen and Funk 2018). Unwritten practice makes it acceptable to take a free right turn when there are no pedestrians or cross-traffic present and it seems obvious that it could be carried out in a nonaggressive way without impeding other cyclists.

Colville-Andersen (2015) calls this a "momentumist" style of cycling, referring to a common scooting momentum that Copenhageners might display when bending the rules a little. This is one of the main differences between Copenhagen and Amsterdam, with Copenhagen requiring subtlety because the system is designed for conformity. Amsterdam, however, is designed for momentum, with more dancing around involved when negotiating busy intersections (Amsterdam has few traffic signals, Copenhagen has many, and so Copenhagen is far more orderly and tidy at intersections).

Reckless cycling is rare in Copenhagen but it can happen. It is highly unusual for a cyclist to fly through an intersection after the green light has changed to red, although from time to time an impatient young man might do just that. Copenhagen cyclists are required to stay on the cycle track while riding their bicycles, and few leave the boundaries. When turning left, cyclists do not merge with motor traffic into a left-hand lane and signal their turn, but instead they perform a "Copenhagen left"—the two-stage box turn.

By comparison, even the most unassuming and law-abiding cyclists in the United States get pulled up. There and in many countries around the world that have ignored cycling for the past century, transportation planners have failed to understand the necessity for cycling as utilitarian transport, and have caused cycling itself to be seen as an act of defiance. This makes cyclists appear reckless since car systems are designed incongruently and are downright hostile to cycling. Lacking the invitation to be on the streets, and the poor condition of infrastructure for cyclists (and pedestrians) makes them seem confrontational rather than welcome. In countries such as the United States this occurs despite the "vehicle codes" that legally allow full access to the streets.

In Copenhagen, car drivers must also be conformists, and this is fundamental to what makes cycling work in the city. At traffic lights showing red, car drivers cannot turn right, and unlike the momentumism of cyclists, this practice is strictly adhered to. When a light changes to green and the traffic moves off, the driver must wait an exceptionally long time by American standards. Being in a car does not make one above the rest, and patience is an obligation as all the cyclists stream by on the driver's right. Only when it is absolutely clear can a driver then make their turn. The cyclists act as a kind of automobile restraint. It is against the law in Denmark to hit a cyclist, and it is almost always legally the driver's fault when collisions occur. Copenhagen's car discipline is profound when compared to the laissez-faire attitude towards and poorly enforced traffic code in the United States and in other countries. But it is also paramount to making Copenhagen a green mobility city.

Cycling in Copenhagen: a quick scan of the literature

Others have also observed the Copenhagen style to varying degrees and many transportation and urban researchers have recognized Copenhagen's allure as a green mobility capital. High cycling rates and car restraint were especially noted in professor of transportation planning Robert Cervero's (1998) *Transit Metropolis: A Global Inquiry*. Using research carried out in the 1990s, Cervero emphasized that bicycling was making "impressive" headway in Copenhagen. Most notably, the city was expropriating curbside car parking for cycle tracks, and had the longest car-free street in Europe at the time (the Strøget in Indre By). Car restraint in the form of high car parking fees, high car taxes, and managed congestion were highlighted by Cervero.

Urban sustainability professor Timothy Beatley's (2000) *Green Urbanism: Learning from European Cities* was enthusiastic about parking removal in Copenhagen's Indre By, extensive pedestrianization, as well as Copenhagen's bicycle mode share, cycle tracks, and integration of bicycling with commuter rail. Researchers at the London School of Economics and Political Science (LSE) highlighted in *Copenhagen: Green Economy Leader* (2014) Copenhagen's relatively low level of car ownership compared to other European cities, emphasizing that low car ownership is an indicator of good environmental performance. On cycling, the LSE described the municipality of Copenhagen's goal of 50 percent bicycle mode share for all work and education trips as unprecedented and observed that over half of all households in the Copenhagen metropolitan region lived within half a kilometer of a commuter rail station, a very suitable distance for cycling.

In *End of Automobile Dependence: How Cities are Moving Beyond Car-Based Planning* (2015) Australian sustainable transportation professors Peter Newman and Jeff Kenworthy praised Copenhagen's inner-city car parking removal policies, and labeled Copenhagen as a "pin-up city" for its high rates of cycling and low car ownership. Aimed largely at a US-based audience, transportation experts John Pucher and Rolf Buehler profiled Copenhagen in *City Cycling* (2012) and asserted that Copenhagen has some of the "world's best" cycling infrastructure and is an exceptionally safe place in which to cycle.

More recently, a cadre of academic researchers based in Copenhagen and in other European cities expanded the global reach and understanding of the bicycle system and bicycle planning in Copenhagen (Gössling, 2013; Carstensen *et al.*, 2015; Nielsen and Skov-Petersen, 2018; Zhao *et al.*, 2018). Malene Freudendal-Pedersen and Jonas Larsen have provided a critical examination of narratives about cycling and discourses about the car in Copenhagen (Freudendal-Pedersen, 2015; Larsen, 2016). Till Koglin (2015a, 2015b) has compared Copenhagen's bicycle and car politics to those in peer cities such as Stockholm, while Copenhagen profiles prominently in *Cycling Cities: The European Experience*, an edited volume on historical perspectives of urban cycling in Europe (Oldenziel *et al.*, 2016).

Outside of academia, Mikael Colville-Anderson (2018) provides a more popularized call to arms for "Copenhagenizing" cities around the world, arguing that a 20–25 percent bicycle mode share is possible in any city in the world. According to Colville-Andersen, it just takes safe, welcoming, connected, and protected bicycle infrastructure and five years of reallocating street space. The European Commission designated Copenhagen the European Green Capital in 2014 largely owing to its cycling rates and ambitions to expand these (European Union, 2013). In Copenhagen, the local Danish not-for-profit public-private partnership State of Green (2016) has celebrated rates of cycling in cycling (although its reporting is largely silent on car restraint). "C40," an international consortium of cities advocating for climate resilience through shared best practice solutions of urban sustainability, has recognized Copenhagen for its exceptional bicycling culture and infrastructure that has resulted in economic and health benefits (but again with no focus on car restraint) (C40, 2016).

Finally, the municipality of Copenhagen has produced exceptionally detailed and updated English translations of key cycling trends and patterns in Copenhagen, and the strong culture of technical expertise informing governance. This literature is even more expansive in Danish-language reports on parking, traffic, and urban planning, many of which inform this book.

The canon of research on Copenhagen's cycling and car restraint is significant and continues to expand, and it reinforces our assessment that Copenhagen's reputation for green mobility is well deserved. The main conclusion drawn from these studies and scholarly publications is that Copenhagen is one of the world's most important bellwethers for sustainable transportation owing to its relatively high rates of cycling and relatively low rates of urban car use.

As we will show later in this volume, this is not a fait accompli nor can it be guaranteed in the future, and so Copenhagen should not rest on its laurels. Yet the data and metrics are indeed impressive, and worth a closer look to see just how inspirational and hopeful Copenhagen might be. Here we will examine the bicycle mode share, or the percentage of ordinary daily trips made by bicycle, and the types of urban form and infrastructure that make these mode shares possible.

Cycling in Copenhagen: data and metrics

Copenhagen's Bicycle Account (City of Copenhagen, 2017a), a biannual traffic survey that also records cyclists' satisfaction with the bicycle system, reports that 29 percent of "all" journeys in Copenhagen are taken by bicycle. "All" trips refers to any daily trip, whether for work or pleasure, both within the municipality of Copenhagen or to its suburbs, and excludes inter-city travel such as train journeys to Aarhus or flying abroad. At 29 percent Copenhagen has one of the highest rates of urban cycling in the world for "all trips", just behind its peer cycling city, Amsterdam, with 32 percent (Kodukula et al., 2018).

The metric indicating all trips is only scratching the surface of Copenhagen's exceptionally high rates of cycling. It is useful to examine Copenhagen's cycling patterns more closely. As displayed in Table 1.1, high rates of everyday cycling in Copenhagen are similarly impressive when broken down to show the purpose of trips, and analyzing where such trips begin and end. For daily commuter journeys (journeys to places of work or education, including ferrying children to school) the rates of cycling hover at around 41 percent and, as mentioned in the Introduction, for commuter journeys originating and ending within the municipal boundaries of Copenhagen, at a striking 62 percent (City of Copenhagen, 2017a). In Amsterdam

TABLE 1.1 Copenhagen bicycle and car mode share by trip purpose

	% of trips by bicycle	% of trips by car
Copenhagen all trips1	29%	34%
Copenhagen commuter trips2	41%	30%
Structures bCopenhagen commuter trips within Copenhagen3	62%	9%
Copenhagen supermarket and street-level shopping trips4	32%	32%
Copenhagen suburbs all trips5	19%	<50%
Denmark all trips6	14%	56%

Sources: 1 City of Copenhagen (2017a); 2 ibid.; 3 ibid.; 4 City of Copenhagen (2015); 5 Capital Region of Denmark (2017); Center for Transport Analytics (2017).

that rate is 48 percent (Harms and Kansen, 2018). Shopping trips within the municipality of Copenhagen are impressive as well, as 32 percent of food shopping trips involve cycling—the same rate as private car drivers (City of Copenhagen, 2015). This is outstanding in the current era given the propensity of North American and, increasingly, European urban dwellers who insist that a car is needed for shopping purposes, especially for families. Furthermore, it is worth noting that in Copenhagen a significant number of households with children—25 percent—use specialized cargo bikes for ferrying children and groceries (ibid.).

As shown in Table 1.2, with the exception of Amsterdam and Utrecht, no other European cities come near Copenhagen's impressive bicycle mode share. Peer green mobility cities in Europe, such as Barcelona, Berlin, London, Paris, Stockholm, and Zurich, have far lower cycling rates by comparison, despite also being celebrated as trendsetters in sustainable transportation (Kodukula et al., 2018). Barcelona and Paris, for example, have introduced aggressive measures designed to rein the use of private cars, but have comparatively low levels of cycling, at 2.3 percent and 3 percent, respectively (City of Barcelona, 2015; Kodukula et al., 2018). London and Stockholm, with congestion pricing schemes that also restrict private car use, have comparatively low rates of cycling (2 percent and 4 percent, respectively) (Koglin, 2015a; Kodukula et al., 2018). Two of Europe's transit city powerhouses, Berlin and Zurich, have better cycling levels (13 percent in Berlin, 8 percent in Zurich) (ibid.; Nash, 2018). Throughout Europe, where approximately 8 percent of all trips are made by bicycle, cities that might have exceptionally high rates of public transit and walking do not match up to Copenhagen in terms of cycling.

We should note that in Groningen, a small city in the northern Netherlands, 61 percent of the population cycle to their place of work or education, and this is

TABLE 1.2 Copenhagen bicycle mode share compared to peer European cities

	% of all trips by bicycle
Copenhagen[1]	29%
Amsterdam[2]	32%
Utrecht[3]	25%
Barcelona[4]	2.3%
Berlin[5]	13%
London[6]	2%
Paris[7]	3%
Stockholm[8]	4%
Zurich[9]	8%

Sources: 1 City of Copenhagen (2017a); 2 Kodukula et al. (2018); 3 Oldenziel et al. (2016); 4 Kodukula et al. (2018); 5 ibid.; 6 Koglin (2015a); 7 Kodukula et al. (2018); 8 ibid.; 9 Nash (2018).

similar to closely with Copenhagen's 62 percent work and education metric. Yet Groningen is a university city similarly to Münster or Heidelberg, both of which have relatively high student cycling shares in Germany, rather than larger urban centers such as Copenhagen or Amsterdam (See Pucher and Buehler, 2012, for comparisons of cycling conditions in small, large, and mega-cities).

Meanwhile, a number of cities in Canada and the United States such as Minneapolis, Portland, San Francisco, Seattle, and Vancouver are renowned for their rates of cycling albeit that these are far lower in comparison to Copenhagen, and hover at or below 5 percent for all commuter trips taken by bicycle (in 2016 Vancouver reported 7 percent; see Table 1.3) (Pucher and Buehler, 2017). In Latin American cities with emergent advocacy promoting cycling, such as Bogotá, Mexico City, and São Paulo, the numbers are just as low, at less than 5 percent according to reports (EcoMobility, 2015a). Bogotá, which introduced the renowned "ciclovía," the largest street cycle scheme of its kind anywhere in the world when the streets are closed to traffic on Sundays and public holidays, is stunningly disappointing at 4 percent despite all the attention the city gets.

Beijing, once recognized as the "bicycle kingdom," with over 60 percent of trips being made by bicycle, has a cycling rate of 12 percent, just behind Berlin (13 percent). As well as North American and European delegations, Beijing planners are looking to Copenhagen for inspiration (Zhao *et al.*, 2018). Osaka, Japan, is also a standout city, with a 25 percent cycling mode share for commuting. Some Indian cities, such as Bangalore and Mumbai, once had higher rates of cycling, but as is the case in much of Africa and the Middle East, cycling remains marginal as rapid motorization has made city streets challenging for cyclists (and pedestrians) (Pucher

TABLE 1.3 Copenhagen bicycle mode share compared to peer cities in North America and Latin America

	% of trips by bicycle
Copenhagen all trips[1]	29%
Bogotá all trips[2]	4%
Mexico City all trips[3]	0.8%
Minneapolis[4]★	3.7%
Portland[5]★	6.3%
San Francisco[6]★	3.9%
São Paulo all trips[7]	1%
Seattle[8]★	3.5%
Vancouver[9]	7%

Sources: 1 City of Copenhagen (2017a); 2 EcoMobility (2015a); 3 ibid.; 4 United States Census Bureau (2016); 5 ibid.; 6 ibid.; 7 Medeiros and Duarte (2013); 8 United States Census Bureau (2016); 9 City of Vancouver (2016).

Note: ★Data represents only bicycle mode share to work

et al. 2005, 2007; Anantharaman, 2016). Australia and New Zealand's cities mirror the low rates of cycling in North America. Few countries (see Table 1.4), despite the predisposition other than the very high rates of car-free and low-income populations in Africa and Asia, have rates of cycling approaching anything near those of Copenhagen (Ouagadougou in Burkina Faso, which has a positive politics towards cycling, is an exception; see Oke *et al.*, 2015).

While Copenhagen's cycling trends are impressive by any standards, the municipality has set itself the goal to increase cycling rates. Most noteworthy is Copenhagen's municipal target of 50 percent bicycle mode share for journeys to places of work and education, whether originating or ending within or outside Copenhagen. Currently the rate is 41 percent. Copenhagen's *2025 Bicycle Strategy*, adopted by the Mayor of Copenhagen and the council, the City of Copenhagen, in 2011 and upheld in 2018, states that the goal is to be able to cycle around Copenhagen without spilling a cup of coffee resting on the handlebars (City of Copenhagen, 2011). Corollary goals are targeting car trips that are currently less than 10 kilometers (6 miles) in length and replacing half of them with cycling (ibid.), and shifting one-third of car trips of between 10 and 15 kilometers (6 and 9 miles) to cycling. Taken together, this should increase work and education trips to 45 percent mode share originating within and outside Copenhagen.

Upgrading the existing bicycle system could boost out the strategy to get to 50 percent. This includes the Plus-Net scheme, such as that which was rolled out on Nørrebrogade, of reducing cycling travel times by improving passability—widening cycle tracks from two abreast to three abreast so that slower cyclists can be safely passed by swifter cyclists (City of Copenhagen, 2011). Copenhagen aims for 80 percent of the city network to be upgraded to Plus-Net by 2025. In 2017 roughly 25 percent of the network incorporated three lanes in each direction, and it should be noted that the widening of cycle tracks is deemed necessary for sociability and to enable parents to ride safely side-by-side with their children.

In addition to the commuting goals, the municipality of Copenhagen seeks to increase the percentage of all trips by bicycle, which currently stands at 29 percent.

TABLE 1.4 Copenhagen bicycle mode share compared to peer cities in Africa, Asia, and Oceania

	% of all trips by bicycle
Copenhagen all trips[1]	29%
Auckland[2]	1.1%
Bangalore[3]	20%
Beijing[4]	12%
Ouagadougou[5]	15%
Melbourne[6]	4%
Osaka[7]	25%

Sources: 1 City of Copenhagen (2017a); 2 Statistics New Zealand (2013); 3 Anantharaman, (2017); 4 Zhao *et al.* (2018); 5 Godard (2013); 6 City of Melbourne (2018); 7 International Transport Forum (2017).

The Copenhagen Municipal Plan (City of Copenhagen, 2015b) calls for a modest increase to 33 percent, or one-third of all trips whether or not they start or end within the city. The municipality's goals for 50 percent of all journeys to a place of work or education trips to be taken by bicycle are more ambitious than its aspirations for all trips. Meanwhile, Copenhagen's globally recognized *2025 Climate Plan*, vague on exact cycling aspirations, seeks the reduction of all car trips to 25 percent, down from current rates, at 34 percent (ibid., 2012).

The municipality of Copenhagen is explicitly seeking to increase cycling and reduce driving and has laid out a Bicycle Path Prioritization Plan and Budget expanding the network until 2025 (City of Copenhagen, 2017c). The city's most recent bicycle count, conducted in 2016 (ibid., 2017a), suggests that the 50 percent target is attainable. Cycling accounts for 62 percent of all work and education trips within Copenhagen, while driving for the same type of trips was just 9 percent. The absolute number of cyclists was up in 2016, and more cyclists were recorded on the harbor and lake bridges, as well in the Indre By, than cars (260,000 bicycles compared to 250,000 cars). Short car trips were declining, with trips of 5 kilometers (3 miles) or less reduced from one-third of all car trips to one-fourth.

The bicycle mode share in Copenhagen's suburbs is also a decent 19 percent (Capital Region of Denmark, 2017). This is noteworthy because in the 120° suburban crescent to Copenhagen's north, west, and southwest, the population density decreases from over 18,000 inhabitants per square mile in the core municipality to between 6,000 and 7,000 inhabitants per square mile in the suburban crescent (according to calculations by Statistics Denmark, 2018a). In the outer exurban edge of Copenhagen, with an even lower density and more open spaces, cycling rates drop to 10 percent, a level that is still higher than in any large American city and which mirrors many of the top cycling cities in Europe. Overall, in Denmark (including Copenhagen), bicycling accounted for 15 percent of all trips in 2016, and 95 percent of Danes considered themselves to be cyclists at least some of the time, even if they owned a car and usually drove (Danish Parliament, 2016). These figures indicate that 26 percent of all Danes cycled at least once a day even if using other modes of transport on the same day, and 80 percent of Danes had a bicycle at their disposal (Olafsson et al., 2016).

TABLE 1.5 Copenhagen goals and aspirations for cycling

	Benchmark	Goal
% of work and education trips originating or ending within or outside Copenhagen[1]	41% (2016)	50% (2025)
% all trips by bicycle[2]	29%	33%
% of cycle tracks with three-person passability[3]	20% (2016)	60% (2020) 80% (2025)
Reduction of car trips[4]	34%	25%

Sources: 1 City of Copenhagen (2016); 2 ibid. (2015); 3 ibid. (2017a); 4 ibid. (2012).

The metrics described above are truly extraordinary when compared to just about anywhere in the world except the equally impressive Netherlands. However, before elaborating further on the key role of urban form and infrastructure, it is useful to clear the air about Copenhagen's topography and weather lest naysayers argue that cycling can thrive only in an environment with a perfect set of conditions such as those found only in Copenhagen and a few other locations.

Clearing the air about sweat and rain: topography and weather in Copenhagen

The topography of Copenhagen is admittedly favorable towards cycling. The Copenhagen region is flat with land elevations near sea level. There are gently rolling hills in the northern suburbs, as well as hills further south on Zealand, but there are only a few steep gradients that cyclists must negotiate. These hills are glacial moraine deposits left when the continental ice sheet retreated northwards 10,000 years ago. The gently undulating forests and hills are an attractive place to cycle for leisure purposes and are easily accessible by bicycle or bicycle-train combination. Where hills are a factor for cycling, Danish cycle planning encourages steepness to be minimized at between 2.5 percent and 4 percent grade, with lengthy inclines for steeper hills (Schonberg quoted in Colville-Andersen, 2012)

However, the wind is far from favorable towards cycling. Copenhagen is located on the Danish island of Zealand and abuts the Øresund, a narrow strait connecting the North Atlantic to the Baltic Sea, and which separates Sweden and Denmark. Nowhere is far from the coast. A strong headwind, which is very common in Copenhagen, can make it feel as though one is cycling uphill. The average winter wind speed is about 22 miles per hour, and headwinds can make it feel like an 8 percent gradient, which is steep for cycling (an incline of less the 5 percent is preferable) (Colville-Andersen, 2018).

Copenhagen is also rainy and cool. It rains almost every other day in Copenhagen (171 days per year on average) and temperatures in the winter can average around freezing. Winters are usually very windy as well. During the winter months in Copenhagen the cold, wet air means that ice can form on cycle tracks overnight. The Plus-Net cycle track system is de-iced or plowed before 7 a.m. on winter weekdays when it freezes or snows. Some 80 percent of Copenhagen's cyclists still cycle throughout the winter (this is partly because the public transit system cannot handle increased capacity. However, as we will discuss later, this might change with the opening of the new Metro City Ring line).

Like most cities around the world, Copenhagen has extreme weather events and very cold or very hot days, but the city also has many good days. Even in the middle of winter the sun can shine brightly and temperatures are pleasant, especially if winds are light. Copenhagen has cool summers, and most homes do not have nor need air conditioning. Rarely are there more than ten days when temperatures rise above 25°C (77°F), making Copenhagen very comfortable for cycling in the summer. But it still rains. Almost every other day.

Colville-Anderson (2018) remarks that instead of overreacting to summer and winter, cities must adapt their bicycle systems to fit the climate. This might mean that shady lanes could be deployed in hot, humid cities in southeastern United States, Asia, or Sub-Saharan Africa. Wet cities can raise cycle tracks out of the gutter, and sink streets for drainage towards the middle rather than edge of the street (which is one benefit of Copenhagen's raised cycle tracks). Winter cities can be designed for snow, and cycle infrastructure can be oriented with sensitivity to prevailing winds. Like humid cities, hot and dry cities could also use shading, as well as minimizing urban heat islands by limiting sidewalks. In many cases it is car space making those cities much hotter.

Buildings can be designed to mitigate the effect of the wind, and cities with a hilly topography can deploy bridges to link places of proximate and similar elevation, use switchbacks, or 'wiggles,' such as the famous route in San Francisco that guides cyclists block-by-block away from steep gradients. When planning levels for bicycles tracks, gradients should be modeled after railroads, which also require very low gradients. In Germany, high-speed rail lines crisscrossing the country transition from tunnels to viaducts in hilly regions. In Atlanta, Georgia, USA, the "belt line," a former railway around the urban core, acts as a catalyst for cycling by providing level connectivity. In Copenhagen, the "bicycle snake," which has a 2 percent gradient, rises from harbor level to an embankment which cyclists previously accessed via steps or had to take a long roundabout detour (see Figures 1.3 and 1.4). Surely cycling, if taken seriously, could undergo similar treatment in hilly cities. There are also many great opportunities for cycling adjacent to railways or former railway routes.

Copenhagen's high rates of winter cycling are not indicative of a perfect climate in the Baltic, but of attention to the details of good cycling conditions. The ice and snow clearance system testifies that conditions can deter cycling if they are not addressed. Moreover, it is standard practice that both new and used bicycles in Copenhagen come with fenders for rain, chain guards, and pedal-generating lights for the darkness of winter (as well as evening cycling). Copenhagen shows that cities, people, and cyclists can adapt to their environment quite easily and with minimal effort.

What makes Copenhagen a standard to aspire to despite the wind, cold, and rain? As the mode share data described above asserts, Copenhagen's bicycle policy is successful and this carries political weight. In the remainder of the chapter we show how Copenhagen approaches the best combination of built environment and infrastructure. In the chapter that follows we outline Copenhagen's cycling history and start to consider how politics has shaped cycling.

Cycling in Copenhagen: bikeability, spatial range, and the compact city

When considering how Copenhagen is the standard to aspire to, we start with the concept of "bikeability," or the urban structures that support cycling (Nielson and Skov-Petersen, 2018). Bikeability is the degree to which the built environment makes it easy to choose to cycle (Colville-Andersen, 2018). Scholarship on

FIGURE 1.3 Image of Copenhagen's "bicycle snake," showing that elevations can be overcome.
Source: Photograph by J. Henderson.

FIGURE 1.4 Image of Copenhagen's "bicycle snake," showing its long, gentle gradient.
Source: Photograph by J. Henderson.

bikeability has blossomed, with Pucher and Buehler (2012) providing a comparative global examination of urban cycling from a US perspective; Horton *et al.* (2007) providing a British context; and there are several edited volumes about bicycles and sustainability spanning European and Asian case studies (Parkin, 2012; Oldenziel *et al.*, 2016).

These and many other works on bicycle cities consider how modern-day cycling is normalized in cities such as Copenhagen or Amsterdam. Other important insights into the bicycle city come from historians examining the first "bicycle boom" in the 1890s and early 1900s when mass utilitarian cycling took hold in Europe and North America (Friss, 2015: Longhurst, 2015; Oldenziel *et al.*, 2016). These historical studies describe bicycle geographies that remain very relevant today.

Friss's (2015) *The Cycling City*, which examined cycling in US cities between 1890 and 1910, develops the concept of a bicycle city from one relying simply on infrastructure to one of a broader bicycle ecosystem overlaying the older walking city and the emerging streetcar city. Invoking the concept of the "annihilation of space through time," cycling was (and remains) up to six times faster than walking with the same exertion, and this expanded the spatial range of the traditional walking city from 1 or 2 miles to 10 or 15 miles radiating outward from the center. This spatial range, or "action radius," as Oldenziel *et al.* (2016) define it, was similar to that of streetcars and buses because at a comfortable pace of 10 miles per hour, cycling was as fast, if not faster, than most streetcars during the 1890s and early 1900s. Streetcars traveled between 9 and 12 miles per hour, but also stopped frequently (Friss, 2015; Oldenziel *et al.*, 2016). This is similar to the pace of many core urban transit systems today. Bicycles also offered easy door-to-door access, could make tighter turns on typically narrow nineteenth-century streets, and were portable and could be carried into buildings and onto trains. Much of this historical bikeability endures today.

In Denmark, bicycle spatial range is analyzed as the average length of a cycling trip, which is approximately between 1 kilometer (0.6 miles) and 5 kilometers (3 miles) (Nielsen and Skov-Petersen, 2018). While these are relatively short distances, there is a huge potential for cycling in cities around the world when considering spatial range. As pointed out in the Introduction, it is notable that in Europe half of all car trips are under 5 kilometers in length, and 30 percent are under 3 kilometers (World Health Organization, 2014). In the United States, 21 percent of car trips are under 1 mile, 46 percent are less than 3 miles, and 60 percent are under 5 miles (United States Department of Transportation, 2017). Such trips take between 15 and 30 minutes on a bicycle, and for this reason there could be a shift towards making many of these trips by bicycle (Parkin *et al.*, 2007; Pucher and Buehler, 2012). In the San Francisco Bay Area, which enjoys a near-perfect climate for year-round cycling, and has vast flat areas where most of the population lives and works, upwards of 72 percent of all car trips are less than 3 miles in length, while 67 percent of all trips are under 5 miles (and yet only 1 percent of trips involve cycling!).

Back in Copenhagen, 44 percent of Copenhageners commute 5 kilometers (3 miles) or less, and 75 percent commute under 10 kilometers (6 miles) (Nielsen and Skov-Petersen, 2018). This helps to explain the remarkable 62 percent bicycle mode share for commuting within the municipality of Copenhagen. Throughout Denmark and including Copenhagen's suburbs, bicycle commuting accounts for between 40 and 60 percent of commuter trips that are between 1 and 5 kilometers (0.5–3 miles) in length, decreasing to less than 10 percent when journeys are more than 15 kilometers (9.3 miles) (Danish Road Directorate, 2017). The practicality and spatial range of commuting by bicycle is especially enhanced by the excellent regional S-train network, and cycling rates increase when they are within suitable cycling range of suburban train stations (Nielsen and Skov-Petersen, 2018). E-bikes, for better or worse, coupled with Copenhagen Capital Region's investment in Cycle Superhighways, extends the spatial range to 10–20 kilometers (6–12.5 miles) or more.

The compact city is also a critical part of the urgent response to climate change and facilitates an easier mode shift from cars. As a proxy for identifying the compact city, population density is important for bikeability because it reflects the number of residences that are in close proximity to each other and to neighborhood retail or other daily services. The higher the density, the more commercial and business services opportunities there are within the spatial range (1–5 kilometers/0.5–3 miles) of cycling (Naess, 2006). The degree of mixed uses within the bicycle's spatial range also matter, especially neighborhood-serving retail outlets and proximity to schools. Nielsen and Skov-Petersen (2018) report that the range of cycling to supermarkets is 3 kilometers (1.86 miles) and to schools it is within 2 kilometers (slightly more than 1 mile) of residences. All things considered, the higher the density, the more likely that the intensity of mixed uses, jobs, and schools should also be higher, with an abundant number of destinations being located within a suitable spatial range of cycling.

When discussing the density and intensity of land use it is worthwhile considering the tout ensemble of the Brokvarterene. If combining all of the bridge district boroughs—Amager, Nørrebro, Østerbro, and Vesterbro, and including the independent municipality of Frederiksburg—the density averages 25,000 persons per square mile. But the Brokvarterene is not just dense, it has a livable density with mixed-use, diverse income (although gentrification pressures are increasing), and it is well designed with attention given to facades and details of street interaction. There are many public spaces and parks, shops, bars, bakeries, cafes, restaurants, and schools. The built environment is dominated by many five- to six-story apartment buildings characterized by a tight perimeter structure with facades following the street, and a semi-private green inner courtyard. Many of the historic building blocks define the street, often with rich architectural detail and with diverse ground floor active uses such as shops, community centers, health clinics, childcare and other family-friendly services.

Districts contemporary to the Brokvarterene exist in cities throughout Europe including Amsterdam, Berlin, Cologne, Frankfurt, Hamburg, Paris, Trieste, Vienna, and Zurich (Sonne, 2017). In Europe, these "reform block" districts were part of a social program that provided air, light, laundries, hygiene, playgrounds, gardens, and a collective approach to community. These built environments are some of the most coherent urbanity of the twentieth century and were a decidedly pro-urban variation of often anti-urban modernism found in twentieth-century planning (Sonne, 2017).

As shown in Table 1.6, beyond the Brokvarterene Copenhagen's density of 18,300 persons per square mile (7,700 persons per square kilometer) is on the low side compared to some peer European green mobility cities, such as Barcelona (41,000 persons per square mile) and Paris (over 54,000 persons per square mile) although Copenhagen's density is higher than that of Amsterdam, with a raw density (that includes water) of approximately 13,000 persons per square mile, London (15,000 persons per square mile), Berlin (11,000 persons per square mile) and Zurich (11,000 persons per square mile).

Nørrebro is the densest district in the Brokvarterene, with 50,000 persons per square mile in 2016, and is denser than the citywide density of Barcelona (41,000 persons per square mile) one of the densest cities in Europe, but less dense than the city of Paris (54,500 persons per square mile). Thus, the opportunity exists to achieve Copenhagen's rates of cycling in many cities around the world if merely allowing for density.

Copenhagen's density range is also comparable to Latin American cities such as Mexico City (16,000 persons per square mile) and São Paulo (23,000 persons per square mile), each of which has a nascent cycling agenda and broader green mobility ambition, and Beijing, the former bicycle kingdom, which still had very dense urban core (64,000 persons per square mile) in 2010, and had 22,000 persons per square mile within the Ring 6 Expressway which circles the main urban conurbation. Bogotá, noteworthy for its green mobility aspirations,

TABLE 1.6 Copenhagen's population density compared to European peers

	Population	Density (persons per square mile)
Copenhagen[1]	611, 822	18,318
Copenhagen bridge districts[2]	383,171	25,545
Amsterdam[3]	851,373	13,000
Barcelona[4]	1,621,537	41,000
Berlin[5]	3,723,914	11,000
London[6]	9,787,426	15,000
Paris[7]	2,206,488	54,000
Zurich[8]	400,028	11,000

Sources: 1 Statistics Denmark (2018a); 2 City of Copenhagen (2017d) and City of Frederiksberg (2016); 3 Netherlands Central Bureau of Statistics (2017); 4 National Statistics Institute (2018); 5 Statistics Office of German States (2018); 6 Office for National Statistics (2018); 7 National Institute of Statistics and Economic Studies (2018); 8 Federal Statistical Office Neuchatel (2015).

TABLE 1.7 Copenhagen's population density compared to Latin American and Asian cities

	Population	Density (persons per square mile)
Copenhagen[1]	611, 822	18,318
Copenhagen bridge districts[2]	383,171	25,545
Beijing (inside the 6th ring road)[3]	11,716,000	22,000
Bogotá[4]	8,080,734	51,800
Mexico City[5]	20,000,000	16,000
São Paulo[6]	12,176,866	23,204

Sources: 1 Statistics Denmark (2018a); 2 City of Copenhagen (2017d) and Frederiksberg (2016); 3 Sixth National Census (2010); 4 Guzman and Bocarejo (2017); 5 INEGI (2018); 6 Metropolitan Region of São Paulo (2018).

has a density of 51,000 persons per square mile, slightly more than that of Nørrebro (Montero, 2018).

Furthermore, Copenhagen has a very similar population density to Los Angeles and San Francisco, two cities aspiring towards green mobility that also enjoy excellent weather conditions that are suitable for year-round cycling. While both the San Francisco Bay Area and Los Angeles are hilly in places, these cities are also incredibly flat across wide swathes of the built-up region, and cycling has great potential. The municipalities of San Francisco and Copenhagen have almost identical densities of around 18,500 and 18,300 persons per square mile, respectively, while core residential districts in each city (for example Market and Octavia in San Francisco and the Brokvarterene in Copenhagen) approach around 27,000 and 25,000 persons per square mile, respectively.

In Los Angeles, which is to the car what Copenhagen is to the bicycle, a significant portion of the city extends from downtown Los Angeles and west-southwest, including East Hollywood, Korea Town, and Westlake, and is almost double's Copenhagen's citywide density with 37,000 persons/mile^2. These Los Angeles neighborhoods are denser than the Brokvarterene, the core of Copenhagen cycling. Meanwhile, the urbanized density of Los Angeles County amounts to 7,000 persons per square mile, and is higher than San Francisco's East Bay and Peninsula suburbs (6,300 persons per square mile), but also higher than the inner suburbs of Copenhagen (6,500 persons per square mile) (Statistics Denmark, 2018a, US Census Bureau, 2015). Most of these Californian urbanized densities are flat and enjoy good year-round cycling weather. It is cold and rainy in Copenhagen for much of the time, yet nearly 20 percent of suburban trips are by cycling.

The Los Angeles and San Francisco Bay Area suburbs have densities comparable to that of Copenhagen but extremely low rates of cycling (under 2 percent regionally), showing that density is necessary but not sufficient for high cycling rates. Los Angeles and other US cities such as New York, Philadelphia, and Chicago have dysfunctional densities because they are characterized by moderate density but also crammed with cars

TABLE 1.8 Copenhagen's population density compared to California's metropolitan areas

	Population	Density (persons per square mile)
Copenhagen[1]	611, 822	18,318
San Francisco[2]	874,228	18,500
Copenhagen bridge districts[3]	383,171	25,545
San Francisco Market and Octavia[4]	30,800	27,000
Los Angeles: Korea Town-Westlake-East Hollywood[5]	320,000	37,311
Greater Copenhagen urbanized area[6]	1,319996	6,554
San Francisco-Oak urbanized area[7]	3,281,000	6,226
Los Angeles urbanized area[8]	12,150,000	6,999

Sources: 1 Statistics Denmark (2018a); 2 California Department of Finance (2017); 3 City of Copenhagen (2017d) and Frederiksberg (2016); 4 Foletta and Henderson (2016); 5 Raimi and Associates et al. (2013); 6 Statistics Denmark (2018a); 7 US Census (2015); 8 ibid.

and built-up environments that are inhospitable to cycling (Chatman, 2008; Edlin, 2010). The high-density parts of California's metropolitan areas are reflective of almost all the high-density areas in the United States, be that Boston, New York (such as Queens), or Seattle, which are high-density cities that are often anti-urban and almost entirely geared towards the car. Hence, in order to achieve the rates of cycling in Copenhagen, density is important, but it is also important to consider that the tout ensemble of the Brokvarterene works very well for cycling, and where cycling rates are high (including in higher-density, high-intensity neighborhoods of US cities like the Mission in San Francisco or Brooklyn in New York), cycling rates are higher as well.

In compact, higher-density developments as envisioned and planned in Europe, North America, and increasingly worldwide, the spatial efficiency of the bicycle is far superior to the spaces of cars. The car simply does not fit. In Europe (where people drive smaller cars than in North America), a car traveling 50 kilometers per hour (31 miles per hour) requires 140 square meters or 15,00 square feet of space, or the size of a large three- or four-bedroom apartment in a typical US city. When parked, that same car requires 20 square meters which is 215 square feet, or equivalent to a large bedroom. Comparatively, a bicycle traveling 15 kilometers per hour (9.3 miles per hour) occupies 5 square meters (53 square feet) and 2 square meters (21.5 square feet) when parked. (Harms and Kansen, 2018). Jan Gehl (2010), using Copenhagen planning specifications, argued that a pair of cycle tracks 6 feet wide running alongside a main street handles 10,000 bicycles per hour. Two car lanes on a one-way street, which is optimal for cars, handles 1,000–2,000 cars per hour at peak times. Meanwhile, one car-parking space equates to ten bicycle-parking spaces. (See City of Copenhagen, 2017a for the 2016 *Bicycle Account* which includes good graphs and charts displaying the allocation of Copenhagen's street space).

Copenhagen's high-density urban tout ensemble, and incredibly high cycling rates show us that considering how space is used is more important than providing raw density estimates. Cycling rates in Copenhagen's lower-density suburbs are higher than in many dense urban centers around the world, and cities with much higher densities (including Paris and parts of Los Angeles) have pitifully low rates of cycling compared to Copenhagen. This tells us that bicycle infrastructure is critical.

Cycling infrastructure in Copenhagen

With its modest high-density and compact urban form, Copenhagen is by all accounts optimally built for cycling, and the city invites people to cycle by providing welcoming infrastructure. Cycle tracks are the mainstay of Copenhagen's cycling infrastructure and get the most attention from outsiders because they are fully separated from cars. For example, in *City Cycling*, Pucher and Buehler (2012) observe that Copenhagen's cycle tracks began to capture the imagination of the United States in 2007 when New York hired consultants from Copenhagen to help to design cycle tracks.

Colville-Andersen (2018) calls Copenhagen's cycle tracks the "gold standard" of cycling infrastructure and surveys in Copenhagen emphasize that cyclists prefer them (Vedel *et al.*, 2017). The provision of cycle tracks makes Copenhagen's cyclists feel like they belong in the city, and with over 230 miles of cycle tracks, they form the backbone of Copenhagen's bicycle infrastructure.

Based on Danish road design standards developed in the late 1930s and early 1940s, cycle tracks are curb-elevated exclusive pathways placed next to the sidewalk. Throughout Copenhagen there is a multi-leveled hierarchy of street elevations with distinctive modes separated by changes in height. The pedestrian level is at the highest elevation, stepping down to the cycling level at the curb, and thence to the car lanes which have the lowest elevations sunk into the middle of the roadway. Buses in Copenhagen share the streets with cars, or in some instances have exclusive busways.

In Copenhagen, the prevailing speed of motor traffic determines the presence of cycle tracks. Based on Danish road design standards, cycle tracks are preferred on roadways with speed limits of between 30 and 40 miles per hour (50–60 kilometers per hour). If cars are parallel-parked along curb, the cycle track is always "parking protected," and is situated on the right side of the parked cars, between the cars and the sidewalk (see Figure 1.5).

Copenhagen's cycle track system differs from many peer cities around the world. In the United States, bicycle lanes (which are delineated by stripes and are level with the rest of the street), are frequently positioned alongside the left-hand side of parallel-parked cars. This places cyclists in the dangerous "door zone" as drivers exiting their cars swing open doors into bike lanes.

In Copenhagen, there are few bicycle lanes, and even fewer like the dangerous "door zone" layout in the United States. When bike lanes or cycle tracks are present in Copenhagen, they are normally placed directly alongside the curb with no adjacent parked cars. However, parking-protected cycle tracks, where present, can pose a door zone problem. In this case the passenger side door can swing into the adjacent cycle

FIGURE 1.5 Image of wide cycle track with excellent passability.
Source: Photograph by J. Henderson.

track. In the municipality of Frederiksberg, which is completely surrounded by Copenhagen and is therefore barely discernable as a separate city, cycle tracks are narrower and door zone conflicts can happen. On Copenhagen's cycle tracks, the passenger side door zone conflict is addressed by widening the cycle track to allow cyclists to steer clear of the door zone, or a buffer is positioned between the cycle track and parked cars. Collisions between cyclists and car doors are rare.

An unwritten rule in Copenhagen is that a pair of cyclists should be able to carry on a conversation while pedaling side by side at a relaxed pace, while a third cyclist should be able to whip around them. For example, Gehl (2010) maintains that in order to facilitate passability cycle tracks should measure between 8.2 feet and 13 feet wide (2.5–4 meters). When calculating the width of cycle tracks (and lanes) it should be taken into account that normal cycling speeds involve subtle zig-zagging or wobbling. If cyclists ride at 12 kilometers per hour (7.5 miles per hour), for example, they zigzag for about 0.2 meters or just over half a foot. At intersections a cyclist who is slowing down is slightly unstable, and should have up to 0.8 meters or 2.5 feet of space (European Union Intelligent Energy Program, 2010). Allowing for cyclists' wobbling and zigzag motions also compensates for error, and so cycling can be safe even in large pelotons.

This, of course, means that more street space must be devoted to cycling. As we shall see in the chapters that follow, widening cycle tracks in any city comes with political conflict over car space, even in Copenhagen. Less apparent, though, is the obstruction caused by utility poles supporting street lights and other urban wiring. In many cities, such as on San Francisco's Market Street, street lighting and utility poles add complications and expense to cycle track proposals. Copenhagen's street lighting system, on the other hand, results in less conflict over space. This is because in Copenhagen street lamps and traffic signals are suspended from cables between buildings, at three to four stories above street level. Copenhagen's dense apartment buildings, closely fronting the street, make it possible to string cables in this manner (Andersen, 2016a). The location of these buildings and municipal willingness to suspend cables between them meant that when Copenhagen's authorities revisited the matter of the city's cycle tracks in the 1980s, it was easier and cheaper to build a new network. Utilities did not have to be relocated as was the case in many emergent cycling cities.

To be sure, cycle tracks do come with caveats and compromises must be made. Bicycles must stay on a cycle track if one is present, and should not wander into the street. Cars can become aggressive if this happens. While cars cannot park on cycle tracks and intrusion is rare, freight delivery is another matter. In Copenhagen, there is an unspoken compromise that freight trucks can straddle cycle tracks but cannot completely block the passage of cyclists, and care must be taken for deliveries during off-peak cycling hours (Andersen, 2016b). Cyclists accept the compromise because freight delivery is necessary for dense urban neighborhoods, but freight operators must ensure that cyclists' safety is not imperiled.

Finally, with few exceptions, all cycle tracks are one way to avoid complications at intersections, which might arguably suggest that cars are accommodated. For example, a cyclist wanting to turn left from a cycle track must perform a two-staged sequence (the "Copenhagen left") which adds some time to the journey, whereas in the United States and in parts of Germany cyclists can turn left in one stage from the center of the street, in a similar manner to a left-turning car. In Amsterdam and the Netherlands, cycle tracks and paths are often bi-directional. Two-way cycle tracks are not found in Copenhagen. In Copenhagen, if a street is a multi-way street, there are two paired cycle tracks, one in each direction. Exceptions include some greenway bike paths, bridges, and some paths on the waterfront and on Amager Strand, a beachfront path. However, since the 1980s bi-directional has been discouraged on city streets.

Cycle tracks are considered necessary when car traffic travels between 30 and 40 miles per hour. When speeds are higher, fully separated cycling paths are required and must be placed as far away as possible from cars. A narrow grassy median, trees, or other buffering features usually separate the path from the adjacent high-speed road. Cycle paths can be bi-directional in certain cases but are normally one way or paired on each side of the roadway.

Copenhagen's regional cycle superhighway system exemplifies the use of separated paths. Cycle superhighways are standardized and signposted cycling routes that extend from the urban core to the exurban edge of the city (Capital Region of Denmark, 2016). These fast, direct, smooth routes invite longer-distance commuting and recreation, with routes ranging from short journeys of 14 kilometers (8.6 miles) to longer ones of 42 kilometers (26 miles). Regionally, the average commute is around 20 kilometers (12 miles) and cycle superhighways (which are also used by electric bicycles) are designed to help to shift suburban driving to cycling. The routes parallel the regional S-train, thus providing complementary access and connectivity between railway stations and suburban residential districts. In Copenhagen, the superhighways also attract reverse commuters to suburban office parks and university campuses. Acknowledging the strong multimodal compatibility of bicycle and trains, bicycles can be taken on the regional S-trains free of charge, and there is generous space inside the rail cars for them.

By paralleling the railways, and because bicycles can be taken on trains free of charge, commuters can cycle for 40 minutes to an hour to work but then opt to take the train back home in the evening. The cycle superhighways are designed to avoid direct contact between yclists and cars traveling at high speed, but sometimes follow highway alignments for directness and to avoid detours. When they do, there is full separation at many intersections, and in the urban core "green waves," timing of traffic signals to average cycling speeds, enables a sustained pedaling speed of 20 kilometers per hour (12 miles per hour) without stopping.

This special attention to the rhythm of cyclists includes considering how repeated stop-and-go is stressful and annoying for cyclists since more exertion and pedal power is required. Uninterrupted flow and service levels help to keep cycling attractive (Gehl, 2010). Green waves increase the travel speeds of average cyclists, reduces sweating from exertion, as well as helping to slow the fastest cyclists, which in turn diffuses aggression and recklessness. It collectivizes and prioritizes the flow and reflects a consensus political culture (see Chapter 2 in this volume).

Intersection approaches are often missing links in the bicycle networks of many cities in the United States, where it is not uncommon to have a bike lane "drop" at an intersection. Often this involves switching from a bike lane to a mixing zone, where right-turning cars can push into the trajectory of cyclists. In congested conditions these mixing zones can become jammed with cars, thus impeding the flow of cyclists who then have to filter awkwardly around stationary vehicles. To reduce the need for such mixing zones, Copenhagen does not allow right turns on red. Indeed, Jan Gehl (2010) states that a "right on red is unthinkable in real bike cities."

Crossing intersections in many US cities can also be challenging as road markings and visible indicators delineating that cyclists belong on the road are missing. In Copenhagen, blue lines with large bicycle stencils are marked across major intersections to show how cyclists should proceed, and to remind motorists that cyclists are likely to be present.

In addition to blue painted markings, cyclists get a head start at many intersections. In Copenhagen, the practice of leading intervals where a special bicycle signal turns green a few seconds before signals turn green for cars has been adopted. These "pre-greens" are joined by stop lines for cars that are set back so that cyclists queue in front of cars and get priority.

Copenhagen's planners avoid reducing the width of cycle tracks at intersections, as these might otherwise provide tempting turning pockets for cars. As mentioned above, intersections are places where cyclists need a little more space to maintain stability. Cyclists are less stable when they slow down compared to the steadiness maintained while traveling at 12 kilometers per hour. When they stop, cyclists need to put a foot to the ground, and intersections should account for cyclists' stability. Wide cycle tracks at intersections account for the nuanced balance and stability cyclists might need and for using their feet to push off again after stopping (European Union Intelligent Energy Program, 2010).

Special bicycle boxes to accommodate "Copenhagen lefts" are also common at Copenhagen's intersections. These indirect left turns involve a two-stage turn with cyclists first crossing the intersection on green, and then performing a fish-hook maneuver on the far side of the intersection, before realigning themselves so that they are perpendicular to the crossing. At major intersections, such as the junction at the Queen Louise Bridge and the lakes, groups of left-turning cyclists cluster together in these spaces (see Figure 1.6). While it is technically legal for a cyclist to continue out of the turn without stopping (if the coast is clear), this rarely occurs.

The complete separation of cyclists and cars is a particular hallmark of Copenhagen and cycle tracks are the backbone of the city's bicycle system, but complete separation is not always necessary. Where neighborhood residential streets are

FIGURE 1.6 Image of cyclists bunching while making a Copenhagen left.
Source: Photograph by J. Henderson.

narrower and have very low traffic volumes, as well as low speed limits of 30 kilometers per hour (18 miles per hour), Danish traffic engineering standards do not recommend separated bicycle infrastructure (Colville-Andersen, 2018). Similar in some ways to bicycle boulevards in the United States or "Woonerfs" in the Netherlands, such streets should be quiet and traffic calmed. Some neighborhood shopping streets are considered welcoming to cyclists even without cycle tracks if the street has low car volumes, 30 kilometer per hour speed limits (18 miles per hour) and high volumes of pedestrians (Vedel *et al.*, 2017).

Shared neighborhood streets can open the entire city to cycling if they are fine-meshed and connected to the mainline cycle tracks, which is how the Brokvarterene is laid out. Traffic-calmed neighborhood streets in the Brokvarterene act as direct short cuts across neighborhoods or connect to trunk line cycle tracks.

It is not considered a good thing if single cycle tracks are isolated and disconnected from other cycle tracks. Systematic connectivity and a cohesive network are key. In Copenhagen, cyclists are confident that wherever they go, they can assume connectivity and cohesiveness. The density of cycle tracks—the distance between parallel routes—is high and fine-meshed.

Copenhagen's fine-meshed network of cycle tracks are designed to make journeys as direct as possible and to keep detours to a minimum. Because bicycle trips usually involve short trips of 1–10 kilometers (0.5–6 miles), most of which are around 2–5 km (1–3 miles) in duration, direct routes are prioritized to make short trips quick, and are not encumbered by detours or meandering pathways that might be suitable for recreation but not for utility cycling. Sometimes such shortcuts warrant contra-flow lanes of cycle tracks, and in certain parts of Copenhagen cyclists are allowed to ride against the flow of cars on one-way, low-speed neighborhood streets. There are few missing links in the network, and all neighborhoods are connected to one another by cycle tracks. At the harbor, multiple bridges are also a major boost to connectivity.

Copenhagen's bicycle bridges played an important part in the upturn in cycling that took place in the 2000s. These fast, direct links include the famous Cykelslangen, or "bicycle snake," that was built in 2014 and by 2016 was being used by 20,700 cyclists daily. It rises from harbor level to an embankment above which previously cyclists reached by taking the steps or a long roundabout detour (see Figures 1.3 and 1.4). With a winding 2 percent gradient the "bicycle snake" is more of an elongated downwards ramp, and the curves slow downhill cyclists for safety. At the harbor edge it connects to Bryggebroen, which was the first of seven bicycle bridges proposed in the early 2000s, and the first harbor crossing to be built in fifty years. Other bridges include the Knippelsbro and Langebro, both of which carry motor vehicles but which have wide cycle tracks to the side of the road. Recently the Inner Harbor bridge opened at the northern end of the main harbor channel, thus relieving some bicycle congestion on the Knippelsbro. This bridge has some poor design issues and cost overruns, but it is envisaged that it will eventually carry 20,000 cyclists daily. A proposed bicycle bridge parallel to the Langebro would similarly relieve the congested cycle tracks on that bridge, while also diverting cyclists off the noisy and polluted H. C. Andersens Boulevard.

A smooth pavement is also a priority. Pavement quality is important to cyclists because poorly maintained cycle tracks can cause annoying vibrations that damage bicycles and can also cause injury, and it is stressful for cyclists having to watch out for bad sidewalks. The municipality of Copenhagen conducts regular surveys of the cycle tracks with a special cycle track measuring vehicle. The sign on the measuring vehicle says "Cykelstimåleren—vi måler for din skyld!" ("Cycle track measurer— we measure for your sake!") As described above, maintaining year-round cycling is yet another high priority in Copenhagen, and the cycle track network is designed to function throughout the winter months owing to an aggressive ice and snow clearance program.

There are numerous subtle measures that make cyclists feel welcome and important, including travel time signs for cars and bikes, digital bicycle counters bearing enthusiastic "thank you" messages, and footrests and railings at some intersections which state: "Hi cyclists, rest your feet here, and thank you for riding in the city." Air pumps and tool stands can be found along the cycle superhigh-ways, special bike ramps are common at curbs and building entrances, and the city ensures that cyclists can pass when construction activity is taking place adjacent to the street. These little things add up quickly, and taken together they make cycling seamless and an easy choice. As a final point, to complete the impressive system of connectivity in Copenhagen's cycling network, public transportation, and espe-cially regional rail, must not be overlooked.

Harmonizing cycling and public transit

Public transit is a keystone making it possible to sustain a large cycling population in Copenhagen. By shifting longer trips from private cars, public transit can help to reduce car traffic, thus making cycling safer and more attractive for short trips in the city (Naess, 2006; Pucher and Buehler, 2012; Nielsen and Skov-Petersen, 2018). On a daily basis some 800,000 public transit trips are taken on buses, trains, and ferries, and Copenhagen's regional trains enable the geographical expansion of the spatial range of cycling.

In Copenhagen, 21 percent of all trips are on public transit compared with 29 percent by cycling, 33 percent by car, and 17 percent on foot (City of Copenhagen, 2016b). Compared to other green mobility cities such as Barcelona, London, Paris, Vienna, or Zurich, which have a public transit mode share of more than 30 percent for all trips, Copenhagen's public transit ridership is less notable, but these cities all have far less cycling (Buehler et al., 2016; Kodukula et al., 2018). Interestingly some US cities, including San Francisco (24 percent), have slightly higher transit mode shares for all trips, but also far lower rates of cycling, and higher rates of car use.

Copenhagen's public transit ridership is slightly higher than that of Amsterdam (17 percent), which, like Copenhagen, has a higher share of cycling. Copenhagen's public transit ridership is within the range of many German cities such as Hamburg (18 percent) or Munich (23 percent), while it is lower than the public transit mode share in Berlin (27 percent) (Buehler et al., 2016; Kodukula et al., 2018).

Copenhagen's high rate of cycling is impressive and inspirational, but Copenhagen's modestly high public transit patronage is also commendable and one of the main reasons why the city is a green mobility capital. But what if cycling and public transportation were synchronized harmoniously? The existence of Copenhagen's regional S-train, together with the relatively high rates of cycling to and from railway stations, exhibits this harmony between cycling and public transit (the S-train is also family-friendly).

The harmonization between Copenhagen's S-trains and cycling enables the geographical expansion of the spatial range for cycling and makes suburban access to S-trains possible without a car. There is ample space on S-train railcars for bicycles (as well as prams, or baby carriages and strollers), and the transportation of bicycles is free of charge, so many cyclists take their bicycle on the transit trip and use it at both ends of their complete journey.

The total length of the S-train network is 169 kilometers (105 miles) and there are eighty-five stations. The trains operate in a radial pattern of about 40 kilometers out of Copenhagen's Central Station, and in 2016 carried 116 million passengers, or between 315,000 and 320,000 passengers per day (Statistics Denmark, 2018b).

The S-train was the organizing template for the historically acclaimed Finger Plan, a 1947 public transit-oriented regional plan, and recognized by Cervero (1998) as a well-run railway. It compares favorably to many similar regional and commuter rail systems in Europe (Danish Transport and Construction Agency, 2016; see also Cervero 1998). The relatively compact urban form proximate to stations meant that by 2012 over half of the Copenhagen metropolitan population lived within 1 kilometer of a railway station, while one-quarter lived within 500 meters of a station (London School of Economics, 2014). In itself this offers tremendous opportunities for cycling. Copenhagen's suburban cycling rate approaches 20 percent of all trips and much of this is oriented around rail and the compact town centers at rail stations (Capital Region of Denmark, 2017). Cycling helps to make Copenhagen's suburban transit-oriented development greener and less car focused (Nielson and Skov-Petersen, 2018).

For residents of denser parts of Copenhagen, such as the Brokvarterene and Indre By, the symbiosis between cycling and the S-train puts urbanites closer to parks, shorelines, meadows and forests, and makes recreation by bicycle possible and affordable. Many Copenhagen residents reverse commute using a combination of cycling and rail, and the emerging cycle superhighway network that parallels the S-train makes it possible to switch easily between modes of transport.

In sum, Copenhagen's S-train is the means by which an expansive regional spatial range for cycling can be achieved. Good modes of public transit such as the S-train facilitate a recalibration of bicycle geographies and our sense about what is possible with the bicycle, and that it is possible to live without a private car.

Although there is a fee for carrying a bicycle on board, the Øresund Railway, which serves Copenhagen (and Copenhagen international airport) as well as Helsingor to the north of Copenhagen, and Lund, Malmö, and Gothenberg, expands Copenhagen's bicycle geographies into southern Sweden.

Conclusion: Copenhagen as a model

Copenhagen is by all accounts optimally built for cycling. Some observers argue that the best bicycle system has already been invented and perfected in Copenhagen, so there is no need to reinvent it (Colville-Andersen, 2018) (observers in Amsterdam will disagree). Yet could not the many elements of Copenhagen's bicycle system simply be implemented in other cities around the world using a "cut and paste" method?

Taking this into consideration, it might be useful to compare the ecosystem of the car city, or the "system of automobility," that involves more than just cars and roads. The system of automobility includes a complex set of zoning and parking requirements, fueling infrastructure, legal and political configurations upholding the right to drive, and social norms framing the car as part of "the good life" (Urry, 2004). Though no purely car city exists, many have pointed to Los Angeles as the quintessential car city, with its freeways and "drive-thru" culture core to that city's identity (Banham, 1971; see also Bottles, 1987; Scott and Soja, 1996).

Today the system of automobility spans the globe and there are elements of Los Angeles everywhere. There are standardized traffic engineering and transportation planning practices (such as parking ratios, lane widths, speed limits, levels of service, and signage) which if absent would make mass motorization and the car culture nearly impossible.

Just as standards for the system of automobility proliferate globally, Copenhagen offers a working model of what a bicycle city looks like. Comparing Copenhagen's cycling system to a smartphone or a well-designed chair, Colville-Andersen (2018) argues that the "Danish design" for cycling infrastructure is universal and that its standards can be applied anywhere with minor modifications to reflect local conditions. The learning curve to use the infrastructure would be short, intuitive, and gentle, and while there is no "one size fits all," there are design standards, cycling etiquette, and assumptions and practices that echo around the world's emerging cycling systems. In Chapter 2 we take a closer look at how this inspiring optimal cycling city came to be, and in subsequent chapters we look at the political pressures that can also be found around the world, and which might undercut the impressive rates of cycling in Copenhagen today.

References

Anantharaman, M. 2016. "Elite and Ethical: The Defensive Distinctions of Middle-Class Bicycling in Bangalore, India." *Journal of Consumer Culture*, 17: 864–886.

Andersen, M. 2016a. *Another Danish Trick for Simpler Streets: Hanging Streetlights*. Boulder, CO:People for Bikes. Available at https://peopleforbikes.org/ (accessed 15 June 2018).

Andersen, M. 2016b. *The Great Copenhagen Loading-zone Compromise*. Boulder, CO, People for Bikes. Available at https://peopleforbikes.org/ (accessed 15 June 2018).

Banham, R. 1971. *Los Angeles: The Architecture of Four Ecologies*. London, Harper Row.

Beatley, T. 2000. *Green Urbanism: Learning from European Cities*. Washington, DC, Island Press.

Bottles, S. 1987. *Los Angeles and the Automobile.* Berkeley, University of California Press.

Buehler, R., Pucher, J., Gerike, R., and Götschi, T. 2017. "Reducing Car Dependence in the Heart of Europe: Lessons from Germany, Austria, and Switzerland." *Transport Reviews,* 37(1): 4–28.

C40 Cities. 2016. *C40 Good Practice Guides: Copenhagen - City of Cyclists.* New York, C40 Cities. Available at www.c40.org/case_studies/c40-good-practice-guides-copenhagen-city-of-cyclists (accessed 31 December 2018).

California Department of Finance. 2017. *Demographics.* Sacramento: California Department of Finance. Available at www.dof.ca.gov/Forecasting/Demographics/ (accessed 14 June 2018).

Capital Region of Denmark (Region Hovedstaden). 2016. *Cycle Super Highways.* Copenhagen, Centre for Regional Development, Capital Region of Denmark.

Capital Region of Denmark (Region Hovedstaden). 2017. *Cycling Report for the Capital Region Hillerode.* Copenhagen, Centre for Regional Development, Capital Region of Denmark.

Carstensen, T. A., Olafsson, A. S., Bech, N. S., Poulsen, T. S., and Zhao, C. 2015. "The Spatio-Temporal Development of Copenhagen's Bicycle Infrastructure 1912–2013." *Geografisk Tidsskrift—Danish Journal of Geography,* 115(2): 142–156.

Center for Transport Analytics. 2017. *The Danish National Travel Survey.* Lyngby, Technical University of Denmark.

Cervero, R. 1998. *The Transit Metropolis: A Global Inquiry.* Washington, DC, Island Press.

Chatman, D. G. 2008. "Deconstructing Development Density: Quality, Quantity and Price Effects on Household Non-Work Travel." *Transportation Research Part A: Policy and Practice,* 42(7): 1008–1030.

City of Barcelona. 2015. *Bicycle Strategy.* Barcelona, City of Barcelona.

City of Copenhagen. 2011. *Good, Better, Best: The City of Copenhagen's Bicycle Strategy 2011–2025.* Copenhagen, Technical and Environmental Administration.

City of Copenhagen. 2012. *CPH 2025 Climate Plan: A Green, Smart, and Carbon Neutral City.* Copenhagen, Technical and Environmental Administration.

City of Copenhagen. 2015a. *Copenhagen City of Cyclists: The Bicycle Count 2014.* Copenhagen, Municipality of Copenhagen.

City of Copenhagen. 2015b. *Copenhagen Municipal Plan 2015: Coherent City.* Copenhagen, Finance Administration: 96.

City of Copenhagen. 2016a. *2016 Annual Parking Report.* Copenhagen, Technical and Environmental Administration.

City of Copenhagen. 2016b. *CPH 2025 Climate Plan: Roadmap 2017–2020.* Copenhagen, Technical and Environmental Administration: 44

City of Copenhagen. 2017a. *Copenhagen City of Cyclists: The Bicycle Count 2016.* Copenhagen, Technology and Environment Administration.

City of Copenhagen. 2017b. *Traffic in Copenhagen: Traffic Figures 2010–2014.* Copenhagen, Technology and Environment Administration.

City of Copenhagen. 2017c. *Cycle Track Priority Plan (2017–2025).* Copenhagen, Technology and Environment Administration.

City of Copenhagen. 2017d. *Statistics and Analysis on Copenhagen: The Population by District and Area, Copenhagen 1 January 2017 and 2016.* Copenhagen, City of Copenhagen.

City of Frederiksberg. 2016. *The City of Frederiksberg.* Frederiksberg, City of Frederiksberg.

City of Melbourne. 2018. *Bicycle Plan.* Melbourne, City of Melbourne. Available at www.melbourne.vic.gov.au/parking-and-transport/cycling/Pages/bicycle-plan.aspx (accessed 12 June 2018).

City of Vancouver. 2016. *Walking and Cycling in Vancouver: 2016 Report Card.* Vancouver, City of Vancouver.

Colville-Andersen, M. 2012. *Danish Bicycle History. Translated from Mette Schønberg. 2009. Traffic and Road, Danish Road History Society.* Copenhagen, Copenhagenize. Available at www.copenhagenize.com/2012/02/danish-bicycle-infrastructure-history.html (accessed 24 June 2018).

Colville-Andersen, M. 2015. *Presentation by Mikael Colville-Andersen to Copenhagenize Masters Class June 23, 24, 25 2015.* Copenhagen, Copenhagenize Design Company.

Colville-Andersen, M. 2018. *Copenhagenize: The Definitive Guide to Global Bicycle Urbanism.* Washington, DC, Island Press.

Danish Parliament. 2016. *Future of Cycling Policy SummitAugust 29, 2016 (Fremtidens Cykelpolitik).* Copenhagen, Government of Denmark.

Danish Road Directorate. 2017. *Driving Forces: Why Is Road Traffic Growing in Denmark?* Copenhagen, Ministry of Transport, Danish Road Directorate.

Danish Transport and Construction Agency. 2016. *Public Transport in Denmark.* Copenhagen, Danish Transport and Construction Agency: 53.

EcoMobility. 2015a. *Bogota, Columbia.* Bonn, EcoMobility Alliance. Available at https://ecomobility.org/alliance/alliance-cities/bogota-colombia/ (accessed 1 June 2018).

EcoMobility. 2015b. *Mexico City, Mexico.* Bonn, EcoMobility Alliance. Available at https://ecomobility.org/alliance/alliance-cities/mexico-city-mexico/ (accessed 1 June 2018).

Eidlin, E. 2010. "What Density Doesn't Tell us about Sprawl." *Access,* 2010 (37): 2-9.

Emanuel, M. 2016. "Copenhagen, Denmark: Branding the Cycling City." In Oldenziel, R., Emanuel, M., de la Bruheze, A., and F. Veraart, eds., *Cycling Cities: The European Experience.* Eindhoven, Foundation for the History of Technology: 77–87.

European Commission 2013. *Copenhagen: European Green Capital 2014.* Luxemburg: European Commission.

European Union Intelligent Energy Program. 2010. *Promoting Cycling for Everyone as a Daily Transport Mode.* Brussels, European Commission.

Federal Statistical Office Neuchatel. 2018. *Population of Zurich.* Available at www.bfs.admin.ch/bfs/en/home.html (accessed 13 June 2018).

Foletta, N. and Henderson, J. 2016. *Low Car(Bon) Communities: Inspiring Car-Free and Car-Lite Urban Futures.* New York, Routledge.

Freudendal-Pedersen, M. 2015. "Whose Commons are Mobilities Spaces? The Case of Copenhagen's Cyclists." *ACME: An International E-Journal for Critical Geographies,* 14(2): 598–621.

Friss, E. 2015. *The Cycling City: Bicycles and Urban America in the 1890s.* Chicago, University of Chicago Press.

Gehl, J. 2010. *Cities for People.* Washington, DC, Island Press.

Godard, X. 2013. *Sustainable Urban Mobility in 'Francophone' Sub-Saharan Africa.* Geneva, United Nations. Available at https://unhabitat.org/wp-content/uploads/2013/06/GRHS.2013.Regional.Francophone.Africa.pdf (accessed 13 June 2018).

Gössling, S. 2013. "Urban Transport Transitions: Copenhagen, City of Cyclists." *Journal of Transport Geography,* 33: 196–206.

Guzman, L. A. and Bocarejo, J. P. 2017. "Urban Form and Spatial Urban Equity in Bogota, Colombia." *Transportation Research Procedia,* 25: 4491–4506.

Harms, L. and Kansen, M. 2018. *Cycling Facts.* The Hague, Netherlands Ministry of Infrastructure and Water Management, National Institute for Transport Policy Analysis.

Horton, D., Rosen, P., and Cox, P. 2007. *Cycling and Society.* Aldershot, Ashgate.

INEGI (National Institute of Statistics). 2018. *Mexico in Facts.* Mexico, National Institute of Statistics. Available at http://en.www.inegi.org.mx (accessed 13 June 2018).

International Transport Forum. 2017. *Bike Share Deployment Strategies in Japan*. Available at www.itf-oecd.org/sites/default/files/docs/bike-share-deployment-strategies-japan_0. pdf (accessed 15 June 2018).

Kodukula, S., Frederic, R., Jansen, U., and Amon, E. 2018. *Living. Moving. Breathing: Ranking of European Cities in Sustainable Transport*. Wuppertal, Wuppertal Institute.

Koglin, T. 2015a. "Vélomobility and the Politics of Transport Planning." *GeoJournal*: 80(4) 569–586.

Koglin, T. 2015b. "Organisation Does Matter: Planning for Cycling in Stockholm and Copenhagen." *Transport Policy*, 39(April): 55–62.

Larsen, J. 2017. "The Making of a Pro-Cycling City: Social Practices and Bicycle Mobilities." *Environment and Planning A*, 49: 876–892.

Larsen, J. and Funk, O. 2018. "Inhabiting Infrastructures: The Case of Cycling in Copenhagen." In M. Freudendal-Pedersen, K. Hartmann-Petersen and E. L. P. Fjalland, eds., *Experiencing Networked Urban Mobilities*. New York, Routledge.

London School of Economics and Political Science (LSE). 2014. *Copenhagen: Green Economy Leader*. London, London School of Economics and Political Science.

Longhurst, J. 2015. *Bike Battles: A History of Sharing the American Road*. Seattle, University of Washington Press.

Medeiros, R. M. and Duarte, F. 2013. "Policy to Promote Bicycle Use or Bicycle to Promote Politicians? Bicycles in the Imagery of Urban Mobility in Brazil." *Urban Planning and Transport Research*, 1: 28–39.

Metropolitan Region of São Paulo. 2018. *Population Density of São Paulo*. São Paulo, Metropolitan Planning Agency. Available at www.emplasa.sp.gov.br/RMSP (accessed 13 June 2018).

Montero, S. 2018. "San Francisco through Bogotá's Eyes: Leveraging Urban Policy Change through the Circulation of Media Objects." *International Journal of Urban and Regional Research*, 42: 751–768.

Naess, P. 2006. *Urban Structure Matters: Residential Location, Car Dependence, and Travel Behavior*. London, Routledge.

Nash, A., Corman, F., and Sauter-Servaes, T. 2018. *A Reassessment of Zurich's Public Transport Priority Program*. Draft Paper Submitted to Transportation Research Board, Zurich.

National Bureau of Statistics. 2010. *Sixth National Population Census of the People's Republic of China*. Beijing, National Bureau of Statistics.

National Institute of Statistics and Economic Studies. 2018. *Population Density of Paris*. Paris, National Institute of Statistics and Economic Studies. Available at www.insee.fr/fr/accueil (accessed 13 June 2018).

National Statistics Institute. 2018. *Population Density of Barcelona 2018*. Madrid, National Statistics Institute. Available at www.ine.es/jaxiT3/Datos.htm?t=2861 (accessed 13 June 2018).

Netherlands Central Bureau of Statistics. 2017. *Population Density of Amsterdam 2017*. The Hague, Netherlands Central Bureau of Statistics. Available at https://statline.cbs.nl/Statweb/publica tion/?DM=SLNL&PA=37230ned&D1=17&D2=39,77&D3= 76,89,102,115,128,141,154,167,180,193,195-196&HDR=G2&STB=G1,T&VW=T (accessed 13 June 2018).

Newman, P. and Kenworthy, J. 2015. *The End of Automobile Dependence: How Cities are Moving Beyond Car Based Planning*. Washington, DC, Island Press.

Nielsen, T. A. S. and Skov-Petersen, H. 2018. "Bikeability: Urban Structures Supporting Cycling. Effects of Local, Urban and Regional Scale Urban Form Factors on Cycling from Home and Workplace Locations in Denmark." *Journal of Transport Geography*, 69(May): 36–44.

Office for National Statistics. 2018. *Population Estimates for the UK, England and Wales, Scotland and Northern Ireland: Mid-2017*. Newport, Office for National Statistics. Available at www.ons.gov.uk/ (accessed 13 June 2018).

Oke, O., Bhalla, K., Love, D. C., and Siddiqui, S. (2015). "Tracking Global Bicycle Ownership Patterns." *Journal of Transport & Health*, 2(4): 490–501.

Olafsson, A. S., Nielsen, T. S., and Carstensen, T. A. 2016. "Cycling in Multimodal Transport Behaviors: Exploring Modality Styles in the Danish Population." *Journal of Transport Geography*, 52(April): 123–130.

Oldenziel, R., Emanuel, M., de la Bruheze, A. A., and Veraart, F. 2016. *Cycling Cities: The European Experience*. Eindhoven, Foundation for the History of Technology.

Parkin, J., ed. 2012. *Transport and Sustainability*, vol. 1, *Cycling and Sustainability*. Bradford, Emerald Insight.

Parkin, J., Ryley, T., and Jones, T. 2007. "Barriers to Cycling: An Exploration of Quantitative Analyses." In D. Horton, P. Rosen, and P. Cox, eds., *Cycling and Society*. Aldershot, Ashgate: 66–82.

Pucher, J. and Buehler, R. 2012. *City Cycling*. Cambridge, MA, MIT Press.

Pucher, J. and Buehler, R. 2017. "Cycling Towards a More Sustainable Transport Future." *Transport Reviews*: 1–6.

Pucher, J., Korattyswaropam, N., Mittal, N., and Ittyerah, N. 2005. "Urban Transport Crisis in India." *Transport Policy*, 12: 185–198.

Pucher, J., Peng, Z.-R., Mittal, N., Zhu, Y., and Korattyswaroopam, N. 2007. "Urban Transport Trends and Policies in China and India: Impacts of Rapid Economic Growth." *Transport Reviews*, 27: 379–410.

Raimi and Associates, 2013. *Health Atlas for the City of Los Angeles*. Los Angeles, Raimi and Associates, City of Los Angeles, and County of Los Angeles.

Scott, A. J.and Soja, E. 1996. *The City: Los Angeles and Urban Theory at the End of the Twentieth Century*. Los Angeles, UCLA Press.

Sonne, W. 2017. *Urbanity and Density in 20th-Century Urban Design*, Berlin, DOM Publishers.

State of Green. 2016. *Sustainable Urban Transportation*. Copenhagen, State of Green: 13.

Statistics Denmark. 2018a. *Population and Elections*. Copenhagen, Statistics Denmark. Available at https://www.dst.dk/en/Statistik/emner/befolkning-og-valg (accessed 3 June 2018).

Statistics Denmark. 2018b. *Metro and S-Train Increase Passengers*. Copenhagen, Statistics Denmark. Available at www.dst.dk/da/Statistik/nyt/NytHtml?cid=24612 (accessed 1 June 2018).

Statistics New Zealand. 2013. *Commuting Patterns in Auckland: Trends from the Census of Population and Dwellings 2006–2013*. Wellington, Government of New Zealand. Available at http://archive.stats.govt.nz/Census/2013-census/profile-and-summary-reports/comm uting-patterns-auckland/commuting-modes.aspx (accessed 3 June 2018).

Statistics Office of German States. 2018. *Population Density of Berlin*. Available at www.sta tistik-berlin-brandenburg.de/pms/2018/18-10-02.pdf (accessed 13 June 2018).

Urry, J. 2004. "The 'System' of Automobility." *Theory, Culture, and Society*, 21(4/5): 25–39.

United States Census Bureau. 2015. *2010 Census Urban and Rural Classification and Urban Area Criteria*. Washington, DC, United States Census Bureau, Geography Division.

United States Census Bureau (American Community Survey). 2016. *Commuting (Journey to Work)*. Washington, DC, United States Census Bureau. Available at www.census.gov/top ics/employment/commuting.html (accessed 15 June 2018).

United States Department of Transportation. 2017. *2017 National Household Travel Survey*. Washington, DC, Federal Highway Administration.

Vedel, S. E., Jacobsen, J. B., and Skov-Petersen, H. 2017. "Bicyclists' Preferences for Route Characteristics and Crowding in Copenhagen: A Choice Experiment Study of Commuters." *Transportation Research Part A: Policy and Practice*, 100(June): 53–64.

World Health Organization (WHO). 2014. *Unlocking New Opportunities: Jobs in Green and Healthy Transport.* Copenhagen, World Health Organization.

Zhao, C., Carstensen, T. A., Nielsen, T. S., and Olaffsson, A. S. 2018. "Bicycle-Friendly Infrastructure Planning in Beijing and Copenhagen: Between Adapting Design Solutions and Learning Local Planning Cultures." *Journal of Transport Geography*, 68: 149–159.

2

A SHORT HISTORY OF CYCLING AND CAR POLITICS IN COPENHAGEN

Introduction

In this chapter we ask: How did Copenhagen sustain high rates of cycling through the twentieth century and into the twenty-first century? We begin by reflecting on how social democracy—a political ideology that includes universal social welfare, human rights, and hefty market regulation of capitalism—may have interacted with cycling and the car. Correlation does not make causation, but the parallels and interconnections between the rise of cycling and the rise of social democracy in Copenhagen should not be ignored. The onset of mass cycling occurred just as social democratic politics was elevated in Denmark. Social democracy was also a legitimate part of Danish politics when the famously high redistributive car tax was adopted. When Denmark established the basic framework for its cycle tracks— safety for cyclists through separation from traffic—there was a social democratic undertone to the consensus and compromise between cyclists and motorists (Carstensen and Ebert, 2012).

The bicycle is not an inherently socially democratic mode of transport, but shared traits between cycling and social democratic principles might have helped cycling to survive during the mid-twentieth century when the car became popular in Denmark and the politics of mobility shifted rightward. Following a period of ambivalence and neglect of cycling, from the late 1960s onward a recharged Left/ Progressive political movement reimagined the bicycle in the 1970s, and put cycling back at the center of Copenhagen's politics of mobility. Cycling rates rebounded slowly at first, and then from the late 1990s and early 2000s a new bicycle renaissance took hold in Copenhagen. This renaissance fused the social democratic politics of cycling in Copenhagen with Neoliberal politics which framed cycling as part of a package of high-value reurbanization and global city competition. Neoliberal framings included attempts to depoliticize the bike, and foreshadowed political tension between Neoliberal, Right/Conservative and Left/ Progressive ambitions for the city. This history lays the foundations for the evolution of a politics of mobility in contemporary Copenhagen.

Social democracy, cycling, and cars

The modern human-pedaled bicycle appeared on North American and European city streets in the 1890s (Friss, 2015). Initially, owing to the relatively high cost of the bicycle, cycling was limited to the elite and it became a status symbol of leisure for the upper class (ibid.; Longhurst, 2015). The bicycle was not an inherently social democratic mode of transport, and in the 1890s the bicycle was an unmistakably capitalist item of conspicuous consumption. The bicycle industry was built on speculation, the rise of mass production, and monopolistic pursuits.

In the 1900s cycling was booming in Copenhagen as low-cost bicycles became indispensable for lower- and middle-class commuters. By the early 1920s 31 percent of trips in Copenhagen were made on bicycles, while 7 percent were by car (Carstensen *et al.*, 2015; Emanuel, 2016). By this time the bicycle had almost completely vanished from the streets in the United States, which was also experiencing a turn-of-century boom. Cycling rates stayed relatively high in many European countries, while cycling started to become more popular in East Asia, including in China and Japan.

As cycling rates climbed in Copenhagen, so too did social democracy, a set of political values that include social welfare, human rights, and hefty market regulation of capitalism. While the bicycle per se was not necessarily social democratic, social democracy was well suited to cycling. Historians of cycling have described a social democratic edge to cycling politics in cities such as Copenhagen and describe how working-class identity, women's emancipation, and social levelling undergirded and subsequently sustained cycling (Carstensen and Ebert, 2012; Oldenziel and Hard, 2013; Emanuel, 2016; Oldenziel *et al.*, 2016).

Social democracy is a political ideology. One of the hallmarks of social democratic ideology is universal publicly funded social welfare—healthcare, education, childcare, maternity leave, pensions, and other social benefits—that is extended to the entire population and publicly financed through the taxation of private income and wealth (including high taxes on private automobiles in Denmark).

The foundational elements of Danish social democracy include the notion that no one should get ahead of anyone else, that everyone is equal, and that bravado, arrogance, or aspirations to be more successful than your neighbors is socially unacceptable (Jenkins, 2011). Historian Bo Lidegaard (2009), in discussing the roots of a "Danish brand" of social democracy, refers to humble acceptance and social responsibility towards those less fortunate, and argues that these values pump-primed Danish culture and society for a universal social democracy. This produced an ideological basis for social solidarity, cooperation, and egalitarianism—albeit in a homogenously white, Danish-only, and mostly Lutheran country with a relatively small population. Subsequent extensions of the social democratic ideal are multicultural, multi-ethnic, multi-national and cosmopolitan, but as Denmark's contemporary immigration politics displays, there are strong nationalist and right-wing populist claims as well.

It is important to distinguish social democracy as ideology that has its roots in the Danish Social Democratic Party (Socialdemokratiet), one of Denmark's oldest, largest, and strongest political parties (and currently the country's leading party). Historically Denmark and other European countries, especially Germany and Sweden, had prominent social democratic parties invoking social democracy, including an active government working for the public good, intervening in the economy to ensure workers' rights and social equity, while contesting privatization and deregulation of the market (Fitzmaurice, 1981; Jones, 1986; Esping-Andersen, 1990; Lidegaard, 2009; Danish Institute for Parties and Democracy, 2018). A core element of these social democratic parties was to strive for social democracy with a focus on workers rather than owners and managers. In a contemporary context, social democracy is synonymous with Left/Progressive ideology in challenging, but not eliminating, private property and celebrating the public good. Social democracy is in theory antithetical to neoliberalism, a market-oriented ideology that privileges deregulated corporations and unfettered, less-regulated private property.

Early social democratic parties were founded with the objective of promoting social democracy, but social democratic ideology transcended political parties (and today social democratic parties might be social democratic in name only) (Fitzmaurice, 1981; Esping-Andersen, 1990). Communists and Socialists, to the left of the Social Democrats, shared social democratic ideals and values but with a more radical critique of capitalism. Along with the Communists and Socialists, det Radikale Venstre, the small urban Social Liberal party, often allied itself to the Social Democratic Party in order to create governing majorities in the Danish parliament; these were instrumental in helping the Social Democrats to implement social democracy (Fitzmaurice, 1981; Lidegaard, 2009). This inter-party cooperation built an expansive universal social welfare state with high taxes but also high-quality public services covering all citizens, albeit only those with Danish residency status (ibid.).

Cycling rates increased in tandem with social democracy, and social democracy was well-suited to cycling. The frugal, simple, human-powered bicycle was a social leveler, universally available at low cost to anyone physically fit enough to pedal. Reflecting humbleness, all class strata in Copenhagen took up cycling (Oldenziel and Hard 2013; Emanuel, 2016). Mass-produced and cheap bicycles became a part of Copenhagen working-class identity and were associated with emancipation and class struggle as working-class bicycle unions were established in Denmark in the early 1900s (as well as in Germany) (Carstensen and Ebert, 2012). Cycling was also associated with the middle class and it was not uncommon to see wealthier Danes cycling. It is under these political conditions of social democracy that rates of cycling in Copenhagen expanded throughout the first half of the twentieth century.

The Brokvarterene was in part a legacy of this early interplay between social democracy and cycling. New reform block social housing was built with funds from trade unions and eventually with state assistance and as non-profit cooperative housing, part of an incremental housing and social program that provided air, light, laundries, hygiene, playgrounds, gardens, and a collective approach to building a community (Larsson and Thomassen, 1991; Sonne, 2017). To be sure, there were also slums in the Brokvarterene, along with speculative housing, and the process of socializing and upgrading affordable housing took decades, and the districts also later underwent gentrification (Hansen *et al.*, 2001; Larsen and Lund Hansen, 2008).

Yet significantly the Brokvarterene and districts like it throughout Europe kept workers in the city and kept them cycling. It was easy to cycle to work, leisure places, church, school, and the shops (Emanuel, 2016). All of the city was within the spatial range of cycling. Later this made it possible to manage easily and comfortably without a car.

The Danish car tax

Threads of social democracy can also be seen in Denmark's famous car tax, a one-off fee that was payable on the purchase of a car and which became an extremely important, albeit unintentional, car restraint policy for Copenhagen (Boge, 2006; Emanuel, 2016). The tax, introduced in 1910 and increased in the 1920s, reflected an early social democratic view of the car. Considered a luxury, the private car was taxed heavily and the resulting revenue steered to the general fund as a progressive and redistributive form of taxation (much of the tax revenue was redistributed to fund social welfare). After the Second World War the car tax was also part of domestic economic policy to stimulate consumption of Danish goods instead of imports as Denmark had no domestic car industry (all cars were imports).

Looking at car ownership between the 1930s and the postwar years, the tax slowed the rate of Danish car ownership when compared to economically similar neighbors. In 1934 there were 34 cars per 1,000 persons in Denmark, but this was on par with Germany, Sweden, and other European neighbors at the time (in the United States, car ownership stood at more than 200 per 1,000 persons, equivalent to Copenhagen's car ownership rate in 2016) (Boge, 2006; Jones, 2008; Oldenziel *et al.*, 2016). The rate of car ownership in Denmark stayed low during the Great Depression and the Second World War—close to the levels of Sub-Saharan Africa and India today—and by the 1960s the rate of car ownership across all of Denmark approached 120 cars per 1,000 persons, much lower than ownership rates in Germany and Sweden and similar to China's per capita car ownership in 2015 (Pineda and Volgel, 2014; International Organization of Vehicle Manufacturers, 2015). Today Denmark's car ownership rate is lower than almost all peer nations.

Unlike Germany, Sweden, the United States, and China (all of which impose lower car taxes), Denmark never had a domestic car manufacturing industry, nor a workforce dependent on manufacturing cars. The Social Democrats and other left-wing parties in Denmark did not have the same political dynamic as their peers in Germany and Sweden, where Social Democrats strongly identified with car manufacturing and grappled with balancing environmental and energy control against the needs of automobile workers (Koglin, 2015a; Koglin and Rye, 2014). In the

United States there is a long history of the automobile industry and trade unions shaping national and state policies on cars, such as car and fuel taxes (Seeley, 1987; Rubenstein, 2001; Jones, 2008; Norton, 2008, Longhurst, 2015; Huber, 2013.)

Many Danes, invoking the doctrine of social democracy, used to be proud of the high car tax and this reflected the Danish brand of a more humble, frugal, but also self-sufficient outlook (City Transport Executive, 2016). The car tax essentially discouraged car ownership among urban dwellers in Copenhagen, and prevented many suburban households from buying a second car (and also compelled them to cycle or use public transit). Under social democratic principles people purchased a car only if it was needed, in keeping with the Danish brand of social democracy and frugality.

The revenue from the Danish car tax helped to finance social democracy. It was not the primary revenue for Denmark by any means, but it was symbolically important, with generous social welfare for all of the population financed through taxation of the elite and of luxury items such as automobiles. For over a century the car tax was a manifestation of social democratic values because it helped to underwrite universal social welfare, metered the rate of car ownership in Denmark, and eventually helped to shape green mobility (Lidegaard, 2009; Transport Advocate and Politician, 2015; Transport Consultant (1), 2016; City Transport Executive (2), 2016, Transport Consultant (2), 2016; Liberal Alliance, 2018).

Safety through separation from traffic

Social democracy, the Danish car tax, and the broader political perspective that the car was a luxury and the bicycle indispensable, meant that the politics of mobility in early twentieth-century Denmark included a politics of protecting cyclists from the car. Between 1900 and the 1920s the construction of sidewalks commenced in Denmark but very little infrastructure was provided for cyclists despite their numbers. Cyclists and motorists increasingly came into conflict on rural roads outside of Copenhagen, where both recreational cycling and recreational motoring clashed. Cars traveling at speed would suddenly come upon a group of socializing cyclists riding at a relaxed pace, thus resulting in frustration for both parties. This scene mirrors contemporary recreational cyclist-car conflicts in rural locations in countries around the globe. In Copenhagen, urban cyclists jostled for space with pedestrians, streetcars, and a smattering of new motorists. By 1915 the Danish Cycling Federation (established in 1905) warned that the car was the enemy to cyclists (Carstensen and Ebert, 2012).

During the 1920s a political compromise gelled. The Danish Cycling Federation conferred with the Federation of Danish Motorists and both advocated for safety through the separation of cyclists and traffic (Carstensen and Ebert, 2012). Copenhagen's first on-street cycle track, built in 1915 on Østerbrogade, was a kind of pilot project that demonstrated the efficacy of such a separation. Separated cycle tracks became normalized in Copenhagen city planning, while paths were laid next to rural roads (Cartsensen et al., 2015).

In contrast to the United States and many other peer nations, the nascent Danish car lobby understood that if cyclists were not to ride in the middle of the road, the cycle tracks needed to be of a high quality. They therefore supported funding for building and maintaining cycle tracks adjacent to roads although tax disputes over the car tax and a proposed cycling tax did arise (Schønberg quoted in Colville-Andersen, 2012; Carstensen and Ebert, 2012). It is also noteworthy that Danish traffic laws made it mandatory for cyclists to stay on the cycle tracks or paths whenever they were available. If there was not, a cyclist could ride in the street with mixed traffic, including all of the streets in the Brokvarterene and greater Copenhagen. Furthermore, the compromise gradually included significant restrictions on car drivers, including forbidding right turns at red traffic lights throughout much of Copenhagen, and very rigid punitive laws against motorists colliding with cyclists.

Oldenziel and de la Bruheze (2011) point out that in many parts of Europe, especially Germany and the Netherlands, separate cycle tracks or paths were promoted not for the benefit of cyclists, but rather for motorists (see also Carstensen and Ebert, 2012). Safety through separation from traffic got cyclists out of the way so that cars could speed up and car drivers would not be liable for hitting cyclists in the road. Within this framework, cyclists were considered a nuisance that ought to get out of the way of cars. This was different in Denmark, where cyclists and motorists were given equal traffic rights, and where the car tax was especially high because cars were considered a luxury to be taxed and regulated.

Historians of European cycling point to separate cycle tracks as being fundamental to sustaining high rates of cycling in cities like Copenhagen (Carstensen and Ebert, 2012; Oldenziel et al., 2016). Both Denmark and the Netherlands, and to a lesser extent Germany, adopted policies that required many roadways to impose safety through separation from traffic (each country also had varying degrees of social democratic politics at the time). When in the later 1930s there were still high rates of road deaths among pedestrians and cyclists, Denmark's road agency established a research program to address road fatalities, which were especially acute in urban areas like Copenhagen where there was dense traffic and housing. In 1939 Denmark required that all counties and municipalities build cycle tracks and pedestrian paths on all roads funded by the national government.

Safety through separation from traffic was built in as Copenhagen expanded and policies like these legitimized and prioritized cycling. This provided a critical and extremely important path-dependent trajectory and footing for cycling rates to remain high, in stark contrast with many peer nations (Carstensen and Ebert, 2012). The contrast is particularly sharp in the United States, where utilitarian cycling was absent for much of the twentieth century.

Contrasting cycling histories in Denmark and the United States

As cities in the United States especially grapple with the car and consider models of green mobility, understanding history also informs us about what needs to be done now. In 1897, the peak of the first bicycle boom, there were five million cyclists in the United States, a not unsubstantial number given that this amounted to about 16 percent of the urban population (Friss, 2015). According to the 1900 US census of industry, there were 312 bicycle manufacturers, with 17,000 workers, producing one million bicycles annually, showing that there was a significant demand for bicycles in the United States during the period (ibid.). In Washington, DC, 18 percent of commuters cycled, and in San Francisco 20 percent cycled. In Minneapolis 44 percent of the population owned a bicycle, and in Chicago one downtown street clocked 211 cyclists per minute, which rivals the busiest cycling streets in modern-day Copenhagen and Amsterdam. Urban cycling was popular enough to impact streetcars, and the private streetcar industry reported declining ridership in some cities due to shifts to cycling in 1897 (ibid.). Then suddenly, cycling rates plummeted in the United States after 1900.

However, private cars did not replace bicycles (ibid.; Longhurst, 2015). During the first two decades of the twentieth century car ownership in cities was limited to the elite and the car was associated with leisure and recreation, not commuting (Jones, 2008). Middle-class urban and suburban car ownership lagged until the mid-1920s, twenty years after cycling rates had crashed. Instead, private railways (streetcars, inter-urbans, and commuter trains) expanded cities of the United States and this impacted streets. With streetcars, horse-drawn carriages and a smattering of new cars, streets became increasingly cluttered and dangerous for cyclists. Many would-be cyclists shifted to public transit (Friss, 2015).

But this is only part of the story. The collapse of cycling was not simply cannibalization by streetcars, nor was it something inherently problematic with the bicycle as a utilitarian urban mode of transport. It was a failure of politics.

More pointedly, because the bicycle was indeed a social leveler it was abandoned by politics. For the urban bourgeoisie in the United States, once the toiling masses took up cycling, cycling lost its cultural appeal as a fashionable social marker and item of conspicuous consumption (Friss, 2015). The premier cycling advocacy organization in the United States, the League of American Wheelman (today's League of American Bicyclists), which was financed by the elite during the 1890s boom, was summarily abandoned by the elite (ibid.; Longhurst, 2015). After 1900, when cycling was taken up by working-class urbanites, there was no movement in the United States to build separate cycle tracks or to defend cyclists' interests in the street.

In Denmark, on the other hand, the Danish Cycling Federation was composed of a mix of members of the bourgeoisie and the urban middle class, but also expressed social democratic politics (Oldenziel and Hard, 2013). The bicycle remained a visible symbol of liberation and social solidarity in these important early decades (Carstensen and Ebert, 2012). By the 1930s Copenhagen was full of cyclists, with 44 percent of all trips made by bicycle in 1935 (Oldenziel et al., 2016).

By contrast, with no political advocates or champions, bicycles were thuggishly pushed off the streets in the United States. The outsized impact of just a few speeding cars, usually driven by the elite, was rapacious and violent with little personal responsibility towards other street users such as pedestrians and cyclists (see Norton, 2008). By the 1920s city streets in the United States had become menacingly hostile to cyclists and with the exception of delivery services cycling almost completely disappeared. There was no national, statewide, or local political lobby advocating for cycling (Friss, 2015). When new traffic codes and regulations were adopted, decisions about street space were ceded to automobile interests. New traffic engineering standards ignored bicycling. Nothing—no street, no sign, no curb, no sidewalk—was designed or engineered with cycling in mind (Longhurst, 2015).

Among the most significant path-dependent impacts of this lack of bicycle politics in the United States was the eventual transfer of the public street curbs to private car parking. This semi-formal entitlement of on-street parking, common in residential neighborhoods and commercial districts, pre-empted other desirable uses of the public right of way, including possible cycle tracks and other important bicycle infrastructure. Aggravating this, almost all cities eventually required off-street car parking. The implementation of this requirement fragmented the public right of way along the curb. Driveways and curb cuts effectively privatized portions of the street that could no longer be used for any other purpose.

Contiguous uninterrupted curbsides, such as those found in much of Copenhagen and many European cities, all but disappeared in the United States. Today ample curbside car parking, and the proliferation of private driveway access, complicate the potential for separate cycle tracks. This usurpation remains one of the most vitriolic and emotional flashpoints in contemporary US bicycle politics (and also, as we will see later, in Copenhagen). Today, in the United States and in many other countries, proposals to repurpose and reimagine curbside parking are extremely controversial in municipal bicycle politics (Henderson, 2013; Longhurst, 2015).

The historic giveaway of public space to private car owners suggests an absence of social democratic principles in how streets are conceptualized in the United States. But this was not due to a lack of social democratic politics. Historically there was a social democratic movement in the United States during the important formative years for cycling (see Kazin, 1989). For reasons that are unclear, it did not include organizing around cycling. The lack of a cohesive social democratic bicycle politics in cycling's formative years and in subsequent years when traffic and parking policies were being drawn up is an important omission in US transport history.

In Copenhagen and throughout Denmark, safety through separation from traffic created a kind of détente on the roads. Cars got the middle and cyclists got the sides, and both tacitly agreed to stay out of each other's way. Parking for cars could be fitted between the center of the street and the curbside cycle track, thereby creating a parking-protected cycle track. When there was not enough space for both parking and the cycle track, historically the city prioritized cycling over car parking.

Safety through separation from traffic defused antagonism between cyclists and car drivers and clarified the rules. It established procedures and guidelines that paired bicycle and car planning. This is very different from what happened in the United States, where violence and antagonism towards cyclists is pervasive in the politics of the car (Culver, 2018). The lack of safe, separated infrastructure in the United States has had a profound impact on cycling, keeping it to a very small portion of travel, and mostly recreational. Deeply hostile antagonism persists between cyclists and motorists, both on the roads and in decisions about transport policy. Perhaps a dose of social democracy can help to bring about change.

Mid-century ambivalence towards cycling and the car

During the Second World War Copenhagen was spared colossal destruction that other parts of war-torn Europe suffered (Lidegaard, 2009). The bicycle city remained in place and cycling rates increased rapidly when rationing was lifted. During the war bicycles were part of the resistance and while there was more recognition and celebration of resistance cycling in Amsterdam (see Jordan, 2013), the bicycle surely had some romanticized cultural cache in postwar Copenhagen (Hong, 2012). At 55 percent of trips in 1949, high rates of cycling in Copenhagen seemed ordinary and secured (Emanuel, 2016).

Despite this high rate of cycling there was ambivalence towards cycling in postwar city planning. For example, the 1947 Finger Plan, a progressive, social democratic map of compact and heavily regulated urbanization that preserved 'green wedges' between radiating fingers centered on S-trains, with Copenhagen as the palm of the hand, mostly ignored cycling (Danish Town Planning Institute, 1993; Cervero, 1998).

The Finger Plan included detailed discussions and maps of future railways, motorways, and major arterial roadways, but a map of a proposed cycle track network or system of pathways was missing. The plan did discuss bicycle parking problems, and recommended keeping bicycles separate from cars as much as possible. Yet when the plan assumed that 65 percent of future homes would be in compact apartment configurations near rail stations—perfect for cycling—there was no mention of cycling in these nodes (Danish Town Planning Institute 1993).

Copenhagen's Municipal General Plan of 1954 did not ignore cycling; rather, it characterized cyclists as a traffic problem (Danish Town Planning Institute, 2004). The plan included suggestions that cyclists should shift to trams because that newly freed-up space would be reallocated to cars. Planners increasingly assumed that there would be many more cars in the future—even with higher car taxes than anywhere else in Europe (Pineda and Vogel, 2014).

Discourses that the car was inevitable in that Danes naturally aspired to car ownership would become common in Denmark just as in Germany and the United States. Increasingly, the car was becoming essentialized—considered natural, expected, and inevitable and inescapable—while cycling was ignored or discouraged. Car ownership rates, still low relative to peer countries, doubled in Denmark in the 1960s, especially in Copenhagen's suburbs where there was increasing political pressure for more roads and parking (Jones, 1986; Pineda and Vogel, 2014).

After the war, Danish economic recovery included raising revenue for the welfare state with high taxation on cars, fuel, and other parts of the car system. The car tax was increased in the 1930s and 1940s, and again in the 1950s. Left/Progressives including Social Democrats, Socialists, and Communists, supported the tax, and during economic emergencies such as the Great Depression wartime austerity and postwar rebuilding, parliamentary members in other parties also supported increasing the car tax. The car tax was popular not just for revenue, but to discourage imports and coordinate Danish spending and resources towards the Danish economy. The tax also made cycling comparatively more affordable.

While discouraging car ownership with the car tax, by the 1950s the Danish Social Democratic Party had joined its Swedish and German counterparts in planning the construction of motorways in Denmark. The Social Democrats believed that motorways would boost economic growth, which in turn would underwrite the expansion of social welfare and fund programs important to Social Democratic voters, some of whom increasingly lived in suburbs and pined for roads (Boge, 2006). The Social Democratic Party was not alone, and all of the main political parties—the Conservatives and the Liberals on the right, and the Social Democrats and the Radicals to the left—endorsed roads and cars (and competed to get votes from the emerging suburban car-owning middle class).

Postwar ambivalence towards the car (discouraging ownership through taxation, while encouraging driving through road building) was infused with new strands of right-wing populism in the 1960s and 1970s. In a major defeat to socially democratic land use planning, a national referendum to strengthen the Finger Plan and compact urbanism was defeated by an anti-tax, anti-left backlash movement (Jones, 1986; Lidegaard, 2009). In the late 1960s this right-populist movement drew in some Social Democrats who had been isolated by their party's leftward tilt during the period (Boge 2006). This conservative, anti-left politics of mobility has remained potent.

Ideas about family, private property, and anti-collectivist politics germinated among households in Copenhagen's suburbs during the political revolts of the 1960s and onwards. In some ways the politics of mobility for the new Danish right wing echoed what Huber (2013) described as a "sunbelt ideology" in the United States centered on privatized cocoons of home, car, and family. Although limited to examining the United States, Huber (ibid.) especially unpacks the relationship between right-wing ideology and gas prices, a proxy for the cost of driving and taxation of cars, and which has been a frequent flashpoint of Danish Right/Conservatives from the 1960s to the present. Right-wing populism argued that the market was apolitical and that it was unfair to have price controls (like the car tax), to redistribute wealth, or for the government to discourage the efforts of entrepreneurial individuals.

Right/Conservative politics, while not necessarily antagonistic towards the bicycle, was antagonistic towards limitations on the car. This stemmed from a kind of privatized social reproduction, emphasizing detached family homes or semi-dettached terraced housing, private property, lower density, dispersal, and auto-mobility (Henderson, 2006, 2013). For the Social Democratic Party this was also an awkward attempt at trying to make the car a component of social democracy. Yet certain aspects of social democracy, such as frugality, cooperation, emphasis on the public over private, and the need for high taxes meant that a social democratic car would be difficult if not impossible. Instead ambivalence towards the car prevailed, whereby pressures to contain the car gave way to pressures to unleash the car.

In this mid-century political ambivalence towards both cycling and the car, Copenhagen was invaded by cars from the suburbs in the 1960s (Carstensen *et al.*, 2015; see also Jan Gehl's (2010) lamentation on Copenhagen at this time). Car traffic doubled on Copenhagen's main streets, while cycling rates were halved between 1960 and 1967. The municipality of Copenhagen planned large parking structures and assumed that motorways would be forthcoming. Images of roads, interchanges and flyovers proliferated in planning for the city, a fifteen-year plan for motorways laid out, and new buildings in parts of Copenhagen were built in anticipation of motorways and more cars (Danish Town Planning Institute, 2004). Copenhagen traffic planners continued to encourage cyclists to switch to public transit in order to make room on the streets for cars, and some cycle tracks and bicycle paths were removed to make room for cars.

The year 1967 was the nadir of cycling in Copenhagen and the mode share for all trips hit a low of 17 percent (car trips were 30 percent) (Oldenziel *et al.*, 2016). Copenhagen was unpleasant and dangerous for cycling, and the highest rates of traffic fatalities in the city's history discouraged more people from cycling (Cartsensen *et al.*, 2015). There was little or no promotion of the bicycle or bicycle infrastructure. Cycling was barely mentioned in a positive way, and many city planners and politicians assumed that the bicycle would eventually disappear (Oldenziel *et al.*, 2016). Road building, the ascendant Danish Right/Conservative politics of mobility, and the essentialization of pro-car policies were expanding. It was obvious that a bicycle city was not pre-ordained. A new politics of the bicycle and the car would be needed if cycling was to remain part of Copenhagen.

Reclaiming bicycle space

A new Left/Progressive politics sparked a new bicycle politics in North America and Europe (Furness, 2010; Carstensen *et al.*, 2015; Emanuel, 2016). In Amsterdam, left-leaning cyclists elevated the bicycle with anti-capitalist bicycle sharing schemes and confrontational bicycle protests advocating that cars be banned from the city center (Wray, 2008; Mapes, 2009; Jordan, 2014; Shorto, 2013; Oldenziel *et al.*, 2016). In Copenhagen, a pro-cycling social movement channeled new political energy from the 1968 student and labor movements that spread through Europe, and explicitly challenged the car (Pineda and Vogel, 2014; Carstensen *et al.*, 2015; Transport Scholar (1), 2016).

Left/Progressive pushback against the car included successful opposition against proposed motorways along Copenhagen's lakes (Pineda and Vogel 2014). In 1968 strong opposition to motorways began, and political leaders recommended better public transit, limiting car parking in the city center, and better traffic management on existing streets (Danish Town Planning Institute, 2004)

In Copenhagen, more radical leftists established Christiania in 1971, declaring the former military base car-free and refurbishing old working-class cargo bikes, kick-starting Copenhagen's cargo bike industry. The bicycle re-emerged as a social democratic symbol with new meanings connected to environmentalism, anti-capitalist politics, "small is beautiful" and locally oriented culture and economy (Carstensen and Ebert, 2012; Oldenziel *et al.*, 2016). The bicycle expressed humbleness, frugality, and social solidarity and was a liberating tool for many Left/Progressives because it enabled car-free living (Transport Scholar (1), 2016). Cycling took on a gendered subtext as well. Cycling was emancipatory for women, and women emancipated by cycling reflected a civilized society rather than a misogynist, belligerent and unequal society arguably displayed in male motorization (Carstensen and Ebert, 2012).

Left/Progressives articulated visions of a non-motorized compact city in the 1970s as membership in the Danish Cycling Federation swelled from 3,000 to 25,000 by 1980. Taking a nod from Amsterdam, Copenhagen hosted car-free Sundays and ideas about new cycling infrastructures, traffic calming, and walkability were proposed. Car-free Sundays reintroduced many Copenhageners to cycling which would have long-term impacts on the return of cycling to Copenhagen (Carstensen and Ebert, 2012). Tired of the city stonewalling and delaying on allocating street space, Copenhagen cycle advocates organized mass protests and drafted their own bicycle plans when the city would not (Oldenziel *et al.*, 2016).

Beyond Left/Progressive politics of mobility, new concerns about energy and resource scarcity captivated Danish politics in the 1970s and 1980s. The 1970s oil crises (1973 Arab oil embargo, 1979 Iranian embargo) were geopolitical events that limited oil supplies and resulted in sudden spikes in gasoline prices. In the United States, panic buying, long lines at gas stations, and pockets of violence followed the abrupt increases in gasoline prices (Yergin, 1990). The crises hit Denmark and the rest of Europe at the same time, but Denmark was one of the most vulnerable oil-dependent nations (Lidegaard, 2009). Oil was not just used for fueling cars, huge oil-fired power plants were part of the national heating and electricity generating scheme. With about 90 percent of Danish energy production based on oil, historian Bo Lidegaard (ibid.) remarks that "if any country was addicted to oil, it was Denmark."

In Copenhagen, Left/Progressives were positioned for the oil crises with their new articulations about cities and nature, and this contributed to redirecting municipal transport policy away from automobility and towards green mobility solutions. Nationally, the governing Social Democratic Party promoted energy conservation and better urban planning, while further to the left ideas about a green decentralized democracy were promoted (Lidegaard, 2009). In order to discourage car and oil consumption, the government increased the national car tax and the value-added tax (VAT), a type of consumption tax levied on goods and services. The national government also cut back road building (Boge, 2006). Motorways already approved and under construction were scaled back as "narrow gauge motorways" with narrower travel lanes and tight medians (ibid.). Mega-projects like the proposed Øresund and Storebælt (Great Belt) bridges were postponed, and motorways around Copenhagen canceled. The oil crisis also kick-started electrification of the national railway, and the main line between Germany and Sweden (via Copenhagen) was electrified, as well as the route between Copenhagen and Odense.

Copenhagen's investment in district heating systems further solidified the response to the oil crisis but also reinforced the higher-density city that works well for cycling. Copenhagen first established district heating in the 1920s but the network was expanded rapidly during the oil crises (City of Copenhagen, 2014). Today Copenhagen has one of the world's largest district heating systems, it is publicly owned, and fits well with the city's high-density cycling tout ensemble (Gerdes, 2013; City of Copenhagen, 2014).

As we have mentioned previously, in Denmark it was arguably easier to elevate cycling and other green mobility solutions because Denmark lacked a powerful domestic car industry such as that found in more highly motorized countries like Germany, Sweden, or the United States (Koglin and Rye, 2014). Sweden developed a globally recognized car industry, but like Denmark had a social democratic political tradition. When the oil crises occurred, Stockholm's cycling rates, which had cratered to 1 percent, did not see increases like those in Copenhagen. Instead, Swedish identity included an economic emphasis on car manufacturing and resulted in public policies that disregarded cycling (ibid.; Koglin, 2015b).

Certainly Stockholm and Sweden had cycling advocacy in the 1970s, but it was unable to cut through the political-economic power of the Swedish car lobby. Copenhagen showed what a city could look like if the political power of car and oil interests were disempowered and their influence on urban space minimized. If we look at the way in which Denmark responded to the oil crisis of the 1970s, we do not see a nation that is resource-scarce but instead a blueprint for a politics of possibilities for how to shift away from fossil fuels and towards green mobility and green energy. Seen in this light, Denmark's limits on energy and resources are not a lesson in disadvantaged or involuntary poverty, but rather an inspiration for how the contemporary global warming crisis might be met head-on with bicycle cities and other public policies such as investment in public transport and renewable energy.

As the oil crises unfolded, Copenhagen also underwent deindustrialization as manufacturing shut down, moved to Jutland or shipped overseas (Winther, 2001). Older industries like textiles, furniture, woodworking, and food processing waned, as did shipbuilding, Copenhagen's largest heavy industry. Economic conditions worsened for the city just as the pro-cycling Left/Progressives organized against motorways and reclaimed city space for cycling. The city collected less tax revenue (especially since high-income professionals were moving to the suburbs) and thus had less to spend on infrastructure.

Some have argued that Copenhagen was spared car-based retrofitting because of accidental-luck of economic and population decline (Koglin and Rye, 2014; Koglin, 2015b; Transport Scholar (2), 2016). Stockholm, Copenhagen's social democratic peer, was doing relatively well in the downturn despite the oil crisis, and invested in motorways and car-centric urban renewal schemes that gutted its urban core. Helsinki, Oslo, and Stockholm all saw new development yet Copenhagen saw none (Larsson and Thomassen, 1991; Andersen and Jørgensen, 1995).

Again, Copenhagen's reclaiming of bicycle space is characterized not as the result of pro-cycling Left/Progressive politics, but as a passive result of exogenous forces of economic troubles. However, as with the oil crisis, this could be examined another way. Instead of just an accident of history, because there was an organized left in Copenhagen, the response to the crises included cycling as an extension of social democratic and progressive values of humbleness, frugality, and thrift. This kind of more virtuous politics of possibilities is worth considering when comparing how many cities around the world responded to economic decline through debt-financed roads and parking as part of the response to economic crises. Deeply depressed cities in the United States, for example, implemented economic development policies that gutted walkable downtowns and public transit and installed large roads and excessive parking.

Copenhagen's economic crisis was a political opportunity to reclaim spaces for cycling, because the bicycle was frugal and required far less money than motorways and urban renewal. In Copenhagen, new cycle tracks were built on many key streets and cycling rates nudged back above 20 percent by the 1980s (Emanuel, 2016). Key elements of the modern bicycle city were introduced including the "Copenhagen left" (a two-stage box turn), blue-shaded lanes to guide cyclists through busy intersections, and advanced stop lines (Transport Scholar (1), 2016). The city officially reintroduced one-way, separated cycle tracks that followed the direction of the car traffic in 1982, and codified the widths, height, and protection of cycle tracks. Parking-protected cycle tracks were preferred where space allowed, and city planners drafted design standards for cycle tracks that estimated cycling volumes based on 1940s levels of cycling (Colville-Andersen, 2015).

Nonetheless, economic decline and restructuring had lasting and unfortunate impacts on Copenhagen. As middle- and upper-class families with children left the city they took with them their income taxes, and Copenhagen transitioned to a city of younger, single students with low incomes, and middle-aged or elderly pensioners with a higher need for social welfare (in Denmark municipalities collect income taxes and funnel revenues into local social welfare programs).

By 1990 Copenhagen was broke, unemployment stood at 15–18 percent, and population in the city dropped below 500,000 (Andersen and Jørgensen, 1995; Andersen and Winther, 2010; Katz and Noring, 2016). Copenhagen took out expensive loans in order to continue social welfare, and tried to create a regional government to rebalance the suburban job and income drain (Andersen and Winther, 2010).

The spiral of deindustrialization in Copenhagen was not met with much sympathy by the Conservative-Liberal coalition (1982–1993), which blocked regional approaches and imposed fiscal austerity on the city. In turn, Copenhagen's social democratic mayor compromised with the national government by embracing neoliberal urban redevelopment schemes (Andersen and Jørgensen, 1995; Hansen et al., 2001). Political fragmentation of the city and its suburbs meant that

Copenhagen competed for higher-wage jobs and residents, and the scheme involved luring higher-income people back to the city in order to improve the city's income tax revenue stream. The Neoliberal vision was upscale inner-city housing around the deindustrialized harbor, concert halls, a new Metro, and finally building the Øresund road and rail link to Sweden, all with the entrepreneurial aim of market-oriented competition with other Nordic and north European cities like Stockholm, Hamburg, and Berlin (ibid.; ibid.; Pineda and Vogel, 2014).

The centerpiece of redevelopment has to be Copenhagen's "havn" (harbor), as it is locally known, which at the time was jointly owned by the municipality and the national government. With 42 kilometers (26 miles) of abandoned docks and deindustrializing waterfront to shape, it was proposed that harbor lands would be cleaned, prepared for new housing or offices, and sold off to developers in order to finance the construction of the proposed Metro.

Cleaning of the polluted harbor commenced in 1992 and front-ended the scheme by making lands adjacent immediately more valuable. By 2014 the municipality boasted that the harbor clean-up had increased adjacent land values by 50–100 percent, thereby stimulating a real estate boom (City of Copenhagen, 2014). Money to finance the construction of the Metro was borrowed up front, and the debt paid down by selling the land opened up by the construction of the Metro, which had become newly appreciated in value. In order to ensure fiscal viability, Copenhagen delayed the sale of lands until the market provided high profit for the city to pay off its debts (Katz and Noring, 2016).

Pre-dating Richard Florida's (2005) "creative class" thesis by a decade or more, Copenhagen was nonetheless pursuing gentrification for the creative class as public policy. As we will see, the bicycle got caught in the crosscurrents of this Neoliberal re-urbanization scheme.

Renaissance: the bicycle boom

Throughout the 1990s and early 2000s, as Copenhagen initiated a neoliberal makeover, cycling rates plateaued at 23 percent in 2000, even after the share of car trips peaked in 1995 at 41 percent, before shrinking to 36 percent in the 2000s (Oldenziel et al., 2016). One of the reasons for which cycling rates stalled in the 20 percent range in the 1990s was that cycling was still poorly understood by the city planning establishment which focused more on urban renewal and also emphasized the recreational side to cycling. Copenhagen's cycling advocates grew frustrated that the city was nibbling around the edges rather than connecting people to where they wanted to go (Transport Scholar (1), 2016).

In 2005, a pivotal year for Copenhagen's bicycle renaissance, bicycling was a top issue in Copenhagen's municipal elections (Koglin, 2015a; Oldenziel et al., 2016). Social Democratic and Social Liberal candidates campaigned on platforms promising to expand the cycling infrastructure and to shift city planning to take utilitarian cycling more seriously. Both left-of-center parties invoked creative class discourses about cycling, citing the need for cycling as a way to lure higher educated but youthful knowledge workers back to the city center and the Brokvarterene (Transport Scholar (2), 2016).

The Social Democrats and Social Liberals prevailed in 2005 and a very pro-cycling figure, Klaus Bondham (who later headed the Danish Cycling Federation) was appointed as head of the city's Technical and Environmental Committee which, after the finance agency, is the most powerful department in the city, and where urban planning and transport matters are deliberated.

Thus began Copenhagen's bicycle renaissance, and between 2006 and 2012 the municipal Social Democrats and Radicals articulated strong pro-cycling visions and adopted new plans and goals, including the aspiration for a bicycle commute mode share of 50 percent (it was 35 percent at the time) (City Transport Planner (1), 2016). Many of the innovations and policies described in Chapter 1 were adopted during this time frame, and Bondham promised to widen cycle tracks in order to address congestion in the Brokvarterene.

During this bicycle renaissance Copenhagen's cycling mode share (for all trips) boomed and surpassed the car's mode share, increasing from 25 percent in 2006 to 32 percent in 2013 (in 2016 it ranged between 29 percent and 32 percent) (Old-enziel et al., 2016). Copenhagen's leaders spoke of making Copenhagen the world's best cycling city and the city undertook the largest wave of expansion since the 1930s when cycle tracks were first built beyond the city center (Cartsensen et al., 2015). Cycle superhighways were proposed to connect suburbs to the core, and the new politics opened ideas for pilot projects, experimentation, and renewed proposals for car restraint.

The city accelerated data collection through traffic counting and surveys, and planned for where cyclists wanted to go, not where traffic engineers thought they should go to be out of the way of cars (Transport Consultant (2), 2016; City Transport Planner (1), 2016). During this time Copenhagen's population decline reversed, and by the mid-2010s the city population had been increasing by 1,000 persons per month, about half of whom were young families having children and who were remaining in the city, and half of whom had migrated from elsewhere.

Especially to North Americans, Copenhagen was seen as leading the way in cycling and Copenhagen became recognized as a global leader and model for addressing global warming. By 2009, the eve of a United Nations Climate Change Summit hosted in Copenhagen, study tours and specialized masterclasses had become popular in Copenhagen. These two- or three-day sessions were aimed at non-profit bicycle advocates, governments officials, and other urban planning-related professionals and organizations who wanted to learn from Copenhagen and then catalyze action back home. A new mini-bicycle boom was emerging in US cities such as Minneapolis, New York, Portland, and San Francisco.

In 2010 Copenhagen hosted the signature "Velo Cities" Conference, which brought global cycling professionals and advocates to convene and corroborate experiences and share ideas. The pro-cycling media in the United States such as *Streetsblog* converged on the city, with catchy short online videos extolling Copenhagen's cycling virtues (Colville-Andersen, 2015). When San Francisco's

City Hall and bicycle advocacy delegation returned home from Velo Cities the City's Board of Supervisors (city council) adopted a resolution targeting a citywide bicycle mode share of 20 percent by the 2020s (Henderson, 2013).

New York's transportation director, Jeanette Sadik-Kahn, said that "Copenhagen inspired" her planning for bicycling and public spaces (Sadik-Khan and Solomonow, 2016). She outlined how New York sent a fact-finding mission to Copenhagen and then exported the first parking-protected bicycle lane to the United States. Copenhagen's urban public space guru, Jan Gehl, was hired to help to redesign Times Square and he convinced the city that cyclists needed barriers (such as parallel-parked cars) to protect them from traffic.

By the 2010s a full bicycle renaissance and restoration was underway. Cycling had become part of a neoliberal rebranding of Copenhagen as a global livable city with a green mobility emphasis. Branding of the bicycle was a healthy business in Copenhagen and Copenhagen deliberately positioned itself as the world's best cycling city while also staging publicity campaigns targeted at Copenhageners to encourage them to cycle (Gössling, 2013). The famous "I Bike CPH" logo was not just a tourist gimmick, but part of a broader cycling awareness and promotion campaign. Cycling was portrayed by the city as positive, healthy, hip, happy, fun, and a playful, humorous discourse was deployed.

A cottage industry of bicycle planning tourism emerged and by 2016 between fifty and seventy delegations were coming to Denmark every year to study Copenhagen's bicycle system (Kabell, 2016). Multi-day bicycle tours, lectures and workshops were offered (for a fee) by the Cycling Embassy of Denmark, an arm of the Danish Cycling Federation. The Cycling Embassy coordinated with People for Bikes, a US-based non-profit organization underwritten by US bicycle retailers and manufacturers, to arrange delegations of American traffic engineers, city hall gadflies, elected officials, and urban planners.

One of the highest-profile study tours took place in 2016 when the Danish Cycling Embassy hosted Anthony Foxx, Secretary of US Department of Transportation (USDOT) in the Obama Administration. Foxx posted his positive experience on the USDOT press briefing site, and stated that "We moved safely through these cities the way so many residents routinely do—on a bike—and we looked at how data and technology are shaping transportation systems for the better" (see Cycling Embassy of Denmark, 2016).

Meanwhile, Copenhagenize, a small consulting firm promoting Copenhagen's bicycle planning which was hired by the municipality for bicycle marketing campaigns, conducted its own masterclasses beginning in 2015 with workshops and bicycle tours for groups of mostly North Americans and Europeans, but also Latin American cycling advocates. The firm also advised cities seeking to increase cycling mode shares, and branched into Belgium, France, and Canada.

The ordinary cyclists of Copenhagen became themselves a global tourist attraction (Freudendal-Pederson, 2015), with American blogs like *Streetsblog* posting photos and videos of the morning bicycle rush hour, and guidebooks like Lonely

Planet and travel writers such as Rick Steves highlighting cycling and encouraging tourists to use bicycles to get around Copenhagen.

Copenhagen itself has become a brand (Gulsrud *et al.*, 2013), an "eco-metropole" with a climate plan and bicycling concept to export, along with industry groups exporting wind energy and other green policies around the world. Copenhagen's Social Democratic Lord Mayor at the time used the city's reputation as a cycling mecca to market Danish green industry, and several Danish green energy firms—Danfoss, HOFOR, Grundfos, Vestas—joined the Lord Mayor on his travels to cities across the world to promote Danish products and green business services. It is not coincidental to see official photos of the Mayor standing next to a bicycle, and for the city's local version of the Chamber of Commerce, via State of Green promote the virtues of cycling (see State of Green, 2016).

Ironically, part of the branding was also the social democratic humbleness, ordinariness, and frugality of the bike: a plain, black, relatively heavy but sturdy and utilitarian bicycle. A clunker bike was good enough to the ordinary Copenhagener, abiding to a social norm of simple, non-exhibitionist bicycles fitted with baskets, fenders, chain guards, and upright handlebars. To further the irony, more expensive versions of the humble bicycle became fashionable and were promoted as new items of conspicuous consumption.

Conclusion: restoration?

Through all of these study tours and workshops, and the promotion of bicycle planning tourism, Copenhagen bicycling was now squarely part of a livable city movement that included urban regeneration and gentrification and was shifting from a decidedly Left/Progressive lifestyle to one of urban consumerism (Carstensen and Ebert, 2012; Emanuel, 2016). A more critical view of bicycle branding suggests that this is just one of a handful of boosterish schemes concocted by Copenhagen's pro-growth coalition, reflecting the neoliberal turn beginning in the mid-1990s, and a shift from welfare planning to entrepreneurial planning (Andersen and Jørgensen, 1995; Baeten *et al.*, 2015; Larsen and Lund Hansen, 2015). Bicycle branding followed in line with redeveloping the industrial waterfront into an urban playground with cultural facilities geared towards tourism and marketing the city to be competitive with Hamburg, Stockholm, and Berlin.

It is important to acknowledge skepticism. The Copenhagenize founder Mikael Colville-Andersen (2015) suggested that green branding may have gone too far. For example, green branding overstates Copenhagen's actual lag in addressing transportation emissions. In this view, too much emphasis on branding has veiled the failure of Copenhagen to reduce transportation emissions from private cars, because even as the city prioritizes cycling, more cars are streaming into the city every day. We will examine this climate gap in the next chapter.

References

Andersen, H. T. and Jørgensen, J. 1995. "Copenhagen." *Cities*, 12: 13–22.

Andersen, H. T. and Winther, L. 2010. "Crisis in the Resurgent City? The Rise of Copenhagen." *International Journal of Urban and Regional Research*, 34: 693–700.

Baeten, G., Berg, L. D., and Lund Hansen, A. 2015. "Introduction: Neoliberalism and Post-Wefare Nordic States in Transition." *Geografiska Annaler Series B: Human Geography*, 97: 209–212.

Boge, K. 2006. *Votes Count but the Number of Seats Decides: A comparative historical case study of 20th century Danish, Swedish and Norwegian road policy.* Ph.D., Oslo, Norwegian School of Management.

Carstensen, A. T. and Ebert, A. K. 2012. "Cycling Cultures in Northern Europe: From 'Golden Age' to 'Renaissance.'" In J. Parkin, ed., *Transport and Sustainability*, vol. 1: *Cycling and Sustainability*. Bradford, Emerald Group: 23–58.

Carstensen, T. A., Olafsson, A. S., Bech, N. M., Poulsen, T. S., and Zhao, C. 2015. "The Spatio-Temporal Development of Copenhagen's Bicycle Infrastructure 1912–2013." *Geografisk Tidsskrift—Danish Journal of Geography*, 115: 142–156.

Cervero, R. 1998. *The Transit Metropolis: A Global Inquiry.* Washington, DC, Island Press.

City of Copenhagen. 2014. *CPH 2025 Climate Plan: Solutions for Sustainable Cities.* Copenhagen, City of Copenhagen and State of Green.

Colville-Andersen, M. 2012. *Danish Bicycle History: Translated from Mette Schønberg (2009) Traffic and Road, Danish Road History Society.* Copenhagen, Copenhagenize. Available at www.copenhagenize.com/2012/02/danish-bicycle-infrastructure-history.html (accessed 24 June 2018).

Colville-Andersen, M. 2015. Presentation by Mikeal Colville-Andersen to Copenhagenize Masterclass, June 23, 24, 25. Copenhagen, Copenhagenize Design Company.

Culver, G. 2018. "Death and the Car: On (Auto)Mobility, Violence, and Injustice." *ACME: An International Journal for Critical Geographies*, 17. Available at www.acme-journal.org/index.php/acme/article/view/1580 (accessed 30 December 2018).

Cycling Embassy of Denmark. 2016. *The United States Secretary of Transportation Wants to Learn Danish Cycling Culture.* Copenhagen, Cycling Embassy of Denmark. Available at www.cycling-embassy.dk/2016/04/20/the-united-states-secretary-of-transportation-wants-to-learn-danish-cycling-culture/ (accessed 24 June 2018).

Danish Institute for Parties and Democracy. 2018. *Danish Political Parties.* Copenhagen, Danish Institute for Parties and Democracy. Available at https://dipd.dk/en/ (accessed 30 June 2018).

Danish Town Planning Institute. 1993. *The Copenhagen Regional Plan/The 1947 Finger Plan.* Copenhagen, Danish Town Planning Institute.

Danish Town Planning Institute. 2004. *Copenhagen General Plan 50 Years Later.* Copenhagen, Danish Town Planning Institute.

Emanuel, M. 2016. "Copenhagen, Denmark: Branding the Cycling City." In R. Oldenziel, M. Emanuel, A. A. de la Bruheze, and F. Veraart, eds., *Cycling Cities: The European Experience* Eindhoven, Foundation for the History of Technology.

Esping-Andersen, G. 1990. "The Three Political Economies of the Welfare State." *International Journal of Sociology*, 20: 92–123.

Fitzmaurice, J. 1981. *Politics in Denmark.* New York, St. Martin's Press.

Florida, R. 2005. *Cities and the Creative Class.* New York, Routledge.

Freudendal-Pedersen, M. 2015. "Whose Commons Are Mobilities Spaces? The Case of Copenhagen's Cyclists." *ACME: An International E-Journal for Critical Geographies*, 14: 598–621.

Friss, E. 2015. *The Cycling City: Bicycles and Urban America in the 1890s.* Chicago, University of Chicago Press.

Furness, Z. 2010. *One Less Car: Bicycling and the Politics of Automobility*. Philadelphia, Temple University Press.

Gehl, J. 2010. *Cities for People*. Washington, DC, Island Press.

Gerdes, J. 2013. "Copenhagen's Ambitious Push to be Carbon Neutral by 2025." *Yale Environment, (*11 April): 360. Available at https://e360.yale.edu/features/copenhagens_am bitious_push_to_be_carbon_neutral_by_2025 (accessed 27 February 2019).

Gössling, S. 2013. "Urban Transport Transitions: Copenhagen, City of Cyclists." *Journal of Transport Geography*, 33: 196–206.

Gössling, S. and Choi, A. S. 2015. "Transport Transitions in Copenhagen: Comparing the Cost of Cars and Bicycles." *Ecological Economics*, 113: 106–113.

Gulsrud, N. M., Gooding, S., and Konijnendijk van den Bosch, C. C. 2013. "Green Space Branding in Denmark in an Era of Neoliberal Governance." *Urban Forestry & Urban Greening*, 12: 330–337.

Hansen, A. L., Andersen, H. T., and Clark, E. 2001. "Creative Copenhagen: Globalization, Urban Governance and Social Change." *European Planning Studies*, 9: 851–869.

Henderson, J. 2006. "Secessionist Automobility: Racism, Anti-Urbanism, and the Spatial Politics of Automobility in Atlanta, Georgia." *International Journal of Urban and Regional Research*, 30: 293–307.

Henderson, J. 2013. *Street Fight: The Politics of Mobility in San Francisco*. Amherst, University of Massachusetts Press.

Hong, N. 2012. *Occupied: Denmark's Adaptation and Resistance to German Occupation 1940–1945*. Copenhagen, Danish Resistance Museum.

Huber, M. T. 2013. *Lifeblood: Oil, Freedom, and the Forces of Capital*. Minneapolis, University of Minnesota Press.

International Organization of Vehicle Manufacturers. 2015. *Worldwide Motorization Rate 2015*. Paris, International Organization of Vehicle Manufacturers. Available at www.oica. net/category/vehicles-in-use/ (accessed 3 June 2018).

Jenkins, R. 2011. *Being Danish: Paradoxes of Identity in Everyday Life*. Copenhagen, Museum Tuculanum Press, University of Copenhagen.

Jones, D. W. 2008. *Mass Motorization and Mass Transit: An American History and Policy Analysis*. Bloomington, Indiana University Press.

Jones, W. G. 1986. *Denmark: A Modern History*. London, Croom Helm.

Jordan, P. 2013. *In the City of Bikes: The Story of the Amsterdam Cyclist*. New York, HarperCollins.

Kabell, M. 2016. *Cycling in Copenhagen*. Presentation to the Future of Cycling in Denmark Summit, 29 August. Copenhagen, Danish Parliament.

Katz, B. and Noring, L. 2016. *The Copenhagen City and Port Development Corporation: A Model for Regenerating Cities*. Washington, DC, Brookings Institution.

Kazin, M. 1989. *Barons of Labor: The San Francisco Building Trades and Union Power in the Progressive Era*. Chicago, University of Illinois Press.

Koglin, T. 2015a. "Vélomobility and the Politics of Transport Planning." *GeoJournal*, 80(4): 569–586.

Koglin, T. 2015. "Organization Does Matter: Planning for Cycling in Stockholm and Copenhagen." *Transport Policy*, 39: 55–62.

Koglin, T. and Rye, T. 2014. "The Marginalization of Bicycling in Modernist Urban Transport Planning." *Journal of Transport & Health*, 1: 214–222.

Larsen, H. G. and Lund Hansen, A. 2008. "Gentrification: Gentle or Traumatic? Urban Renewal Policies and Socioeconomic Transformation in Copenhagen." *Urban Studies*, 45: 2429–2448.

Larsen, H. G. and Lund Hansen, A. 2015. "Commidifying the Danish Housing Commons." *Geografiska Annaler: Series B, Human Geography*, 97: 263–274.

Larsson, B. and Thomassen, O. 1991. "Urban Planning in Denmark." In T. Hall, ed. *Planning and Urban Growth in the Nordic Countries*. 1st edn. London and New York, E & FN Spon: 6–59

Liberal Alliance. 2018. "Old Charges on New Cars." *Liberal Alliance News Magazine*(spring). Copenhagen, Liberal Alliance.

Lidegaard, B. 2009. *A Short History of Denmark in the Twentieth Century*. Copenhagen, Gyldendal.

Longhurst, J. 2015. *Bike Battles: A History of Sharing the American Road*. Seattle, University of Washington Press.

Mapes, J. 2009. *Pedaling Revolution: How Cyclists Are Changing American Cities*. Corvallis, Oregon State University Press.

Norton, P. D. 2008. *Fighting Traffic: The Dawn of the Motor Age in the American City*. Cambridge, MA, MIT Press.

Oldenziel, R. and de la Bruheze, A. A. 2011. "Contested Spaces: Bicycle Lanes in Urban Europe, 1900–1995." *Transfers: Interdisciplinary Review of Mobility Studies*: 29–49.

Oldenziel, R. and Hard, Mikael. 2013. *Consumers, Tinkerers, Rebels: The People Who Shaped Europe*. Basingstoke, Palgrave Macmillan.

Oldenziel, R. Emanuel, M., de la Bruheze, A. A., and Veraart, F., eds. 2016. *Cycling Cities: The European Experience*. Eindhoven, Foundation for the History of Technology.

Pineda, A. F. V. and Vogel, Nina. 2014. "Transitioning to a Low Carbon Society? The Case of Urban Transportation and Urban Form in Copenhagen since 1947." *Transfers: Interdisciplinary Review of Mobility Studies*, 4: 4–22.

Rubenstein, J. M. 2001. *Making and Selling Cars: Innovation and Change in the US Automotive Industry*. Baltimore, Johns Hopkins University Press.

Sadik-Khan, J. and Solomonow, S. 2016. *Street Fight: Handbook for an Urban Revolution*. New York, Viking.

Seely, B. 1987. *Building the American Highway System: Engineers as Policy Makers*. Philadelphia, Temple University Press.

Shorto, R. 2013. *Amsterdam: A History of the World's Most Liberal City*. New York, Vintage Books.

Sonne, W. 2017. *Urbanity and Density in 20th-Century Urban Design*. Berlin, DOM Publishers.

State of Green. 2016. *Sustainable Transportation*. Copenhagen, State of Green. Available at https://stateofgreen.com/en/publications/sustainable-urban-transportation/(accessed   27 February 2019).

Winther, L. 2001. "The Economic Geographies of Manufacturing in Greater Copenhagen: Space, Evolution, and Process Variety." *Urban Studies*, 38: 1423–1443.

Wray, H. J. 2008. *Pedal Power: The Quiet Rise of the Bicycle in American Public Life*. Boulder, CO, Paradigm Publishers.

Yergin, D. 1990. *The Prize: The Epic Quest for Oil, Money, and Power*. New York, Simon and Schuster.

Interviews held by Jason Henderson from 2015 to 2017 with transportation experts in Copenhagen

City Transport Executive (1), 2016.
City Transport Executive (2), 2016.
City Transport Planner (1), 2016.

Transport Advocate and Politician, 2016.
Transport Consultant (1), 2016.
Transport Consultant (2), 2016.
Transport Scholar (1), 2016.
Transport Scholar (2), 2016.

3

"SOMETHING IS ROTTEN IN THE STATE OF DENMARK"

The plateauing of cycling and the rise in car ownership in Copenhagen

"Something is rotten in the state of Denmark"

We began this book by highlighting Copenhagen's impressive cycling metrics. Twenty-nine percent of all daily trips is an extraordinary level of cycling compared to just about every other big city in the world except Amsterdam or Utrecht. The rate of cycling rises to 32 percent of all trips if just considering Copenhageners' daily travel and subtracting trips by Copenhagen's suburbanites. Even more remarkably, 62 percent of all work and school trips starting and ending within Copenhagen are made by bicycle, and bicycles outnumber cars in the urban core (City of Copenhagen, 2017a).

Likewise, only 9 percent of work and school trips starting or ending within Copenhagen are by car, and short car trips (under 5 kilometers, or roughly 3 miles) have declined from one-third of all car trips to one-fourth (City of Copenhagen, 2017a). These impressive metrics are the result of providing a world-class bicycle system and a politics of mobility in Copenhagen that deliberately prioritized cycling while also implementing car restraint policies.

Yet there is dissatisfaction and concern that cycling rates in Copenhagen are plateauing while car use is increasing. Copenhagen's "50 percent goal"—whereby 50 percent of all work and school trips with an origin or destination in Copenhagen and including trips taken by both Copenhageners and non-Copenhageners—might not be realized. Exacerbating this, between 2000 and 2014 car ownership in Copenhagen grew by 29 percent, to a rate of 195 cars per 1,000 persons, or roughly 110,000 cars in the city (City of Copenhagen, 2017b).

Car ownership remains relatively low in Copenhagen compared to peer cities, but between 2014 and 2016 11,000 additional cars were registered in the city, bringing the total to 121,000 (by comparison there were 675,000 bicycles in Copenhagen in 2016) (City of Copenhagen, 2017a). It has yet to be determined whether the rate of increase in car ownership might level off but car ownership in Copenhagen's suburbs and throughout the rest of Denmark is certainly growing, and this suburban car use is putting demands on streets and parking within Copenhagen.

While city Copenhageners like to cycle, suburban Copenhageners like to drive. To borrow from Shakespeare's *Hamlet* (which was set in Helsingor, 45 kilometers (28 miles) north of Copenhagen), "Something is rotten in the state of Denmark." With cycling possibly plateauing and car ownership rising, Copenhagen's reputation as an iconic cycling city might be compromised.

There are several trends that point to unsettling anxiety in this green mobility capital. First, on the outskirts of Copenhagen there are a lot of cars, and fewer bicycles. On an average weekday in 2016, 560,000 cars crossed Copenhagen's municipal border, compared to only 53,000 cyclists (City of Copenhagen, 2017a). Bicycle trips across the municipal boundary are falling, dropping by 16 percent between 2014 and 2016 (ibid.).

Second, in the urban core, cyclists outnumber cars, but this is a recent occurrence, and not guaranteed to stay that way, especially if, as some predict, many cyclists shift to the new Metro City Ring line when it opens in 2019 (Kabell, 2015). While not necessarily adding car trips, if cyclists shift to the Metro, the balance on the streets in the core might shift to cars. Motorists, having the slim majority on the streets, might demand more space, or at least resist further reallocations of space; this situation is already happening.

Third, there is growing concern in Copenhagen about parents driving children to school. This is a relatively new trend in Denmark, but chauffeuring children by car has reduced rates of walking and cycling to school in the United States, where less than 13 percent of children walked or cycled to school in 2010, and the trend was downward (Safe Routes to School, n.d.; Tumlin, 2012).

In Copenhagen, 19 percent of children are transported to school by private car, while 28 percent cycle and 40 percent walk (City of Copenhagen, 2017a). Historically most children walked or cycled because in Copenhagen and much of Denmark public schools are close to residences. However, with the opening of new private schools there has been more driving, and throughout Denmark more children are being driven and cycling to schools is declining (Danish Parliament, 2016). This is profoundly worrisome because when parents cycle, this behavior usually gets passed on to children who then grow up cycling, and continue to cycle as adults (Carstensen and Ebert, 2012). If parents drive, motorization might accompany their children into adulthood.

Fourth, there is also anxiety over bicycle ownership. While in Copenhagen bicycle ownership has remained stable (there is more than one bicycle for every Copenhagener) it is not growing, however, and beyond the municipality there is unease (City of Copenhagen, 2017a). *Politikin,* (one of Denmark's most widely read newspapers) referring to Statistic Denmark data in April 2018, reported that nationally fewer bicycles were being sold, while car sales were rising (Astrup, 2018a). In 2017 bicycle sales in Denmark dropped by 25 percent compared to 2015.

Fifth, new challenges have emerged. While Denmark has banned the transportation network company Uber, these car-based services have managed to nudge their way into the city and Uber logos are likely to appear on smartphone screens, albeit illegally. Meanwhile, Copenhagen is grappling with the proliferation of electric bicycles and the implications for the cycle tracks of higher speed, heavier, bulkier bicycles. Tensions are rising on the cycle tracks and debate over speed limits for e-bikes and access to the cycle track are starting to unfold. In 2018 electric scooters started to be used in Copenhagen and became a new flashpoint on the cycle tracks.

Copenhagen's cycle tracks, especially in the Brokvarterene, Frederiksberg, and at key choke points like the Queen Louise Bridge, have serious capacity issues. Copenhagen built the cycle tracks, the cyclists came, and now there is concern that overcrowding and long queues on the cycle tracks are bringing frustration and discouragement (Transport Scholar (2), 2016; City Transport Planner (1), 2016). A corresponding capacity problem is lack of adequate bicycle parking. In surveys about cycling conditions in Copenhagen, crowding and parking are the main complaints.

In one way the capacity problem reflects success, because the infrastructure attracted more cycling, and in terms of absolute volume there are more cyclists in Copenhagen than ever (City of Copenhagen, 2017a). Too many cyclists is not a bad problem to have. Copenhagen's solution to the capacity problem has been to widen the cycle tracks and build more bicycle parking, both of which necessitate reallocating street space away from cars. However, as we shall see in this chapter and the remainder of the book, this is easier said than done. There is a pushback against further car restraint, and a politics of mobility that is making Copenhagen more, not less, like peer cities in the US and around the world.

The plateauing of cycling in Copenhagen

Copenhagen's goal for 50 percent of all trips to places of work and education to be taken by bicycle was adopted by the Copenhagen city council in 2005. Shortly after local elections which resulted in big wins for cyclists, the city's aspiration was to meet that goal by 2015 (City of Copenhagen, 2011). Between 2012 and 2014 it appeared that the city was well on its way to meeting the 50 percent target, as cycling rates for work and places of education shot up from 36 percent to 45 percent in just two years (see Figure 3.1). The spike in 2014 was a sudden 25 percent increase (ibid., 2015a). Was it a trend or an anomaly?

Before 2014 the bicycle mode share for work and education trips starting or ending in Copenhagen hovered at around 36 percent and was only slightly nudging upward, so the spike to 45 percent was enthusiastically welcomed as showing the results of coordinated investment and promotion of cycling (City Transport Planner (1), 2016). Yet, as suddenly as it spiked, the rate of cycling for work and education dropped at the next count in 2016, to 41 percent. It has not increased since and instead may have plateaued. While still showing good results, the current

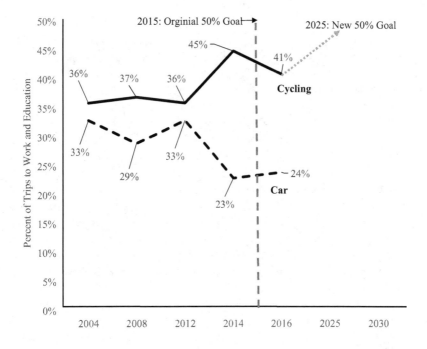

FIGURE 3.1 Graph showing Copenhagen's cycling rates for work and education, 2005–2025 (with car mode share for comparison).
Source: Graph by J. Henderson.

rate is 9 percentage points lower than the 50 percent goal. Furthermore, the 50 percent goal was pushed out to 2025, suggesting that local planners and politicians might be losing momentum for expanding cycling rates compared to previous ambitions (City of Copenhagen 2017a; City Transport Planner (1), 2016).

A dissection of cycling conditions in Copenhagen between 2012 and 2016 is warranted. To some extent climate might be a factor. Between 2008 and 2010 winters were cooler and wetter, and cycling rates stayed "normal" holding for all other factors. Between 2012 and 2015 winters were milder, and this might have induced more people to cycle (City of Copenhagen, 2011; Gössling, 2013; Colville-Andersen, 2015). It was also during this period that regular ice and snow clearance of cycle tracks began, which also helped to maintain year-round cycling. As discussed previously, climate factors in Copenhagen's cycling rates, albeit not significantly. At current rates, 75 percent of Copenhagen's cyclists continue to cycle year round (ibid., 2018). So what else might have elevated cycling to 45 percent of all trips to places of work and education in 2014?

It became much quicker to cycle in the city center than to drive (or to take the bus). During construction of the Metro City Ring, which commenced in 2012–2013, many streets were dug up and blocked off, citywide car traffic speeds dropped to a crawl as multiple construction sites impeded streets. A 3.5 kilometer (2 mile) trip by bus could take 18 minutes, 15 minutes by car but just 9 minutes by bicycle (Movia, 2016). Copenhagen's bus operator, Movia, observed that ridership on city buses was declining because cycling had become more attractive as buses were slowed down by the traffic (ibid.). The city provided passage for cyclists through construction zones while cars were diverted (Colville-Andersen, 2015).

The construction impacts on car traffic and cycling showed what a more robust car restraint policy might look like in Copenhagen. The construction of the Metro City Ring might have metered car volumes in the city. The City Ring includes 17 future stations that encircle Indre By and the Brokvarterene (see Figure 7.1). During construction, which involved boring new rail tunnels and shafts for each new station, these became choke points. The city also renovated the district heating system adding to traffic delays for cars. In the time that cycling spiked up to 45 percent of all work and school trips, car commuting declined, from 33 percent in 2004 to 23 percent in 2014. This was despite the fact that the population and number of jobs had increased in Copenhagen (Oldenziel *et al.*, 2016). Considering the disruption to car traffic due to Metro and utility construction, car restraint works. Traffic jams and the removal of car parking spaces increased cycling.

Coinciding with the spike in cycling and the drop in driving, the actual situation on the cycle tracks, for many cyclists, was becoming uncomfortable. There was not enough space for 35,000 additional cyclists in 2014. As we will see below, there is still not enough capacity even after the brief peak passed and cycling rates settled at around 41 percent. Copenhagen, having successfully wooed more cyclists, had nowhere to put them. Planners pondering this overcrowding and capacity problem now recognize that if Copenhagen is to meet the 50 percent goal, more street space must be reallocated to cycling, thus taking away more space from cars. If Copenhagen is to remain a green mobility capital, this reallocation must also ensure that the buses are not stuck in traffic, meaning even more space must come from cars.

Overcrowding and capacity for cycling in Copenhagen

Overcrowding on the cycle tracks is a major impediment for sustaining Copenhagen's remarkably high levels of cycling and also impedes efforts to take Copenhagen's cycling to the 50 percent goal (Danish Parliament, 2016; Transport Consultant (2), 2016).

City surveys and interviews with cyclists reveal concerns about the flow on the cycle tracks, lack of visibility because of crowding, and animosity towards new, unaware and inexperienced cyclists (including many tourists cycling in the urban core) (Freudendal-Pedersen, 2015). The main cycle tracks have become so crowded at rush hour that it takes longer to cycle through the city, especially at intersections with signals. To cycle on Copenhagen's main arterial cycle tracks is less relaxing and requires more vigilance to avoid crashes. More recklessness is emerging, and with wide cargo bicycles, new e-bikes, and more recently e-scooters, impatience can be further accentuated (Larsen and Funk, 2018).

In one survey of 4,000 cyclists, conducted at the height of the bicycle renaissance in 2011, 82 percent of respondents stated that crowding was unsafe and 69 percent stated that arguments and negative interaction was increasing on the cycle tracks (Vedel *et al.*, 2017). The same survey also showed that many cyclists in Copenhagen were willing to cycle up to 1 kilometer or more out of their way just to avoid highly congested cycle tracks, as well as cycle tracks that had frequent stop-and-go traffic. That was in 2011, well before the 2014 peak. In terms of crowding and congestion, things may have worsened.

A survey of cyclists in late 2016 pointed to frustration with the famous "Copenhagen left" as crowding escalates at intersections where left-turning cyclists bunch up and conflict with through traffic (Larsen and Funk, 2018). As impatience builds, and turns are delayed by two lights, there might be an increase in the gentle rule-breaking described in Chapter 1, but at a rate that could create more friction on the cycle tracks.

Congestion on cycle tracks is thought to discourage new cyclists (City Transport Planner (1), 2016). New residents moving to Copenhagen from within Denmark or from abroad are not cycling due to intimidation by the crowding, and even the "bicycle curious," or people with a pre-disposed desire to try cycling, might be discouraged (City Transport Planners (2 and 3), 2016).

Parents have become concerned about children being caught up in such congestion and therefore avoid cycling, meaning that many children are now growing up without cycling being part of their daily routine (City Transport Planners (2 and

FIGURE 3.2 View from the handlebars during morning rush hour in Copenhagen. Source: Photograph by J. Henderson.

3), 2016). Children are obviously smaller than adult cyclists, and their inexperience of moving through traffic and lack of visibility makes parents nervous. When a child outgrows being transported in a cargo bike, they might stop cycling in the city (Freudendal-Pedersen, 2015). Meanwhile, parents that do use cargo bicycles to transport their children also face hostility from other cyclists when the cycle tracks are narrow and passing is difficult.

The growth in cargo bicycles has exacerbated crowding and congestion. Some have pointed out an "SUV-ification" of cargo bicycles, as they get larger and wider, taking up more space on the cycle tracks, sometimes with rude or entitled parents plodding among other cyclists (Colville-Andersen, 2015). Nonetheless, cargo bicycles are appropriate in a green mobility city. One-third of cargo bicycle owners use them instead of cars and 26 percent of Copenhagen families with children possess cargo bicycles (City of Copenhagen, 2015a).

Cargo bicycles amount to roughly 7 percent of all the bicycles in the city (approximately 40,000 in number) (Colville-Andersen, 2015). Clearly, if this trend continues and bicycle flows are to be maintained, Copenhagen will need to widen the cycle tracks. Furthermore, if, as some cycling advocates are suggesting, cargo bicycles start to be used for urban freight delivery, spaces will have to be identified and allocated with this usage in mind.

Cyclists express frustration with cars too. When the cycle tracks are crowded, and adjacent street space is also filled with cars, it becomes apparent that there is an inequitable crowding of bicycles. In Copenhagen, 7 percent of citywide road space is taken up by cycle tracks and paths, while road space for cars amounts to 54 percent, and for car parking, another 12 percent. Pedestrian space amounts to 26 percent (City of Copenhagen, 2017c). Where cycle tracks or bike lanes have not been widened, cyclists are hemmed in while cars have more room. Where there are no cycle tracks, such as neighborhood streets, cyclists are forced to hug the curb and this inadvertently privileges the car and makes bicycles subordinate, very much like in most of the United States (Larsen and Funk, 2018). Although only 9 percent of trips within Copenhagen are by car, cars consume more than half of road space. Colville-Andersen (2018) calls this the "arrogance of space."

If the city intends to increase cycling rates, more shared spaces with mixed traffic will be required, which also necessitates slowing cars and constraining their movement. Surveys in Copenhagen's bicycle accounts suggest that roughly 74 percent of cyclists feel safe most of the time, but the goal of the city is that 90 percent of cyclists should feel safe by 2025 (City of Copenhagen, 2016a).

Pollution from cars is a source of nuisance to Copenhagen cyclists, especially the particulate pollution that occurs during the winter months (Vedel et al., 2017). In the 2011 survey of cyclists, 47 percent of respondents said that they were extremely dissatisfied with pollution on their commute (ibid.). Being near cars and sharing space with polluting cars and trucks meant that many cyclists were willing to go out of their way to avoid some pollution hot spots, especially if there was a suitable detour with green surroundings.

Bicycle-bus conflicts are less frequent but they do happen (Danish Parliament, 2016). Most of Copenhagen's trunk line cycle tracks are adjacent to the main bus routes. Conflict arises at bus stops, many of which are placed on the sidewalk side of the cycle track, requiring passengers to board and disembark directly onto the cycle track (see Figure 3.3). Cyclists are supposed to come to a complete stop and give right of way to people descending from buses, and this adds to the stop-and-go crowding along some of the cycle tracks.

Seniors and disabled bus riders are especially vulnerable at these kinds of bus stops, and some collisions between elderly people and cyclists have occurred (Colville-Andersen, 2015). Owing to these unfortunate incidents, and in order to accommodate cyclists and bus passengers, new boarding islands have been fitted between the cycle track and the street, so there are fewer conflicts and cyclist flow can be maintained. However, this means that more car space must be removed in order to provide sufficient space for both bicycles and buses.

Inadequate bicycle parking tops the frustrations of many Copenhageners who cycle, especially in shopping areas and at train stations (City of Copenhagen, 2017a). Surveys of cyclists show that parking is rated poorly and is annoying and inconvenient. (Pucher and Buehler 2012; Danish Parliament, 2016). Copenhagen zoning rules require bicycle parking to be incorporated in new buildings, with 2.5 bicycle parking stalls needed for each 100 square meters of residential space, and

FIGURE 3.3 Cycle track and bus stop conflict.
Source: Photo by J. Henderson.

0.5 stalls for each employee in office developments (Gössling, 2013). But home and work are not the parking problem. Dissatisfaction is high at train stations and in shopping areas.

It is essential to be able to park in convenient locations, and to lock and unlock bicycles easily in a cycling city like Copenhagen, and Copenhageners have adapted to the inadequate parking situation through leaning bicycles against buildings or "fly parking" by using a kickstand and frame lock (Larsen and Funk, 2018). This explains the abundance of bicycles parked on sidewalks and along the entire length of buildings. In a windy city like Copenhagen, however, one often sees bicycles being knocked down like dominoes and owners have to dig out their bicycles from beneath piles of fallen bicycles.

Over 32 percent of shopping trips in Copenhagen are made by bicycle, and at supermarkets parking can be frustrating because of the lack of sufficient parking spaces (City of Copenhagen, 2015a). This especially holds for cargo bikes which, being more expensive, are more likely to be chained to a rack or signpost instead of fly parked. Sometimes parked cargo bicycles are awkwardly placed and take up more room to the chagrin of other cyclists and shopkeepers. Often parking gets less attention than the more glamorous cycle tracks, but Copenhagen shows that inadequate bicycle parking can dampen enthusiasm.

Meanwhile, to stem the tide of plateauing or declining rates of cycling in Copenhagen's suburbs, hope hinges on the cycle superhighway concept, which has its own set of challenges and shortcomings. On paper, new cycle superhighways are standardized, fast, direct, routes with smooth sidewalks and intersection priority for cyclists (or underpasses). The cycle superhighways are regional but most center on Copenhagen. The aim is to make crossing from the outer suburbs into Copenhagen more palpable for longer distance commuting (20 kilometers or 12.5 miles), including reverse commuting from Copenhagen outward, and to provide Copenhageners with easy cycling access to regional open spaces and recreation areas. Emerging pedal-assisted e-bike commuters and aligning the routes parallel to the S-train system have the potential of shifting car trips to bicycle trips on the outskirts of Copenhagen. Within Copenhagen, the superhighways overlap with the Plus-Net routes targeted for capacity expansion (Bennetsen and Overgaard Magekund, 2018).

Implementation of the cycle superhighways has been mixed and somewhat underwhelming. Due to the lack of adequate funding and buy-in from certain suburbs, planning for cycle superhighways has been slow, with three routes introduced between 2009 and 2016, and the rollout of half-a-dozen more in 2018. For the most part each route was overlaid on existing cycle tracks and cycle paths that were stitched together into a single cohesive route. Some of the routes are very winding and indirect (Bennetsen and Overgaard Magekund, 2018). Many routes also had been positioned next to very busy and noticeably loud highways, thus making the cycling experience less than enjoyable.

Political fragmentation is part of the problem since the administrative region leading the planning of cycle superhighways—known as Capital Region—does not have road building authority. This is the preserve of the national government in respect of motorways, and of the municipalities for most other roads and streets (Regional Planner (1), 2016). There are twenty-nine municipalities in the Copenhagen region and between twenty-three and twenty-five have in some way cooperated with the superhighway scheme. It is not compulsory to participate, and some Neoliberal and Right/Conservative suburbs such as Hillerod in the north of Copenhagen have balked at the superhighways.

In the meantime, some cycle superhighways have sections that are actually a long way from railway stations and so lack complementarity. As we will elaborate in Chapter 6, other segments of cycle superhighways have been undermined by car parking. In Copenhagen's outer Vanlose neighborhood, one cycle superhighway route forces cyclists to mix with cars because a separate path was not built. Car parking trumped cycle superhighways (Bennetsen and Overgaard Magekund, 2018). In Frederiksberg, which has high rates of cycling and is part of the Bork-varterene, political conservatives and some merchants opposed routing a cycle superhighway on the Gammel Kongevej (Old King Road) which is a direct, and popular radial route across the municipality, but which would require reduction in car space to accommodate both cycle superhighway standards and improved bus stops (City Transport Planner (4), 2016).

The promotion of cycle superhighways by the Capital Region Administration is in conjunction with promotion of pedal-assisted electric bicycles, and faster non-pedal "throttle" electric bicycles. E-bikes of both kinds make up 10 percent of cycling sales in Denmark (City Transport Planner (4), 2016) and are causing more conflict on the cycle tracks. The average speed of a bicyclist in Copenhagen is 15.8 kilometers per hour, or 10 miles per hour. Some of the newer throttle bikes approach speeds of 45 kilometers per hour (28 miles per hour), which is too fast for a cycle track. Yet the Danish National Government legalized the use of these faster e-bikes in 2018 even as the Danish Cycling Federation and City of Copenhagen warned that this was an invitation for conflict and undercuts safety for other cyclists (Politiken, 2018).

Similar conflicts emerged when e-scooters were also allowed on cycle tracks on a pilot basis in January 2019. Adding fast-moving e-bikes and scooters to the already tense situation on the cycle tracks, while perhaps accommodating non-car commuting for some, could further complicate capacity and crowding. As a result, some of Copenhagen's cycling advocates have suggested that e-bikes are folly (Colville-Andersen, 2014). Others have criticized e-scooters (Astrup, 2018b)

In response to all of the concerns outlined above, and after surveying over 10,000 cyclists in 2016, Copenhagen's new Bicycle Plan recognizes that over-crowding and capacity are a major problem (City Transport Planner (4), 2016). If the 50 percent goal becomes reality, the absolute number of bicycles in the city center will rise from 180,000 in 2010 to 240,000 by 2025. Major expansion of existing infrastructure—the Plus-Net concept described in Chapter 1—will be needed to accommodate 60,000 more bicycles daily within the city center. Streets will need to be reallocated for bikes and space must be taken from that used by cars.

Some observers have suggested that the reallocation of streets might not be needed if, when the new Metro City Ring line opens, many cyclists shift to the new Metro, thus relieving pressure on the cycle tracks, and relieving the political pressure to take more car space (City Transport Planner (1), 2016; Kabell, 2015). Taken a step further, the dynamic between the opening of the Metro City Ring and the spaces on the surface above could shift in favor of accommodating more cars, in an ironic repeat of the twentieth-century allocation of space in cities such as London, New York, and Paris, where the relocation of the railway underground freed up space for cars above the ground.

For other observers the core issue remains that the car has too much "right to the city" and this is what is creating the crowding and lack of space for cyclists (Freudendale-Pedersen, 2015). Even if the Metro does siphon off cyclists the city center should be made less welcoming to cars, and perhaps entirely car free. Although there are new car owners within Copenhagen, and especially in newer developments in the harbor area, the political pressure for car space is also coming from the outskirts and suburbs of Copenhagen, which have taken a separate mobility path from Copenhagen. We therefore need to assess car ownership trends, and to break down the politics of this separate suburban car-oriented trajectory.

Suburbs that love to drive: car ownership in Copenhagen and Denmark

The municipality of Copenhagen, with a population of approximately 614,000 in 2018, is the political, economic, and cultural capital of Denmark and has been the focal point of this examination of the iconic cycling city (Statistics Denmark, 2018b). Private car ownership in Copenhagen remains relatively low compared to peer cities but the city is experiencing a profound increase in car ownership, with new data from Statistics Denmark showing that in 2018 there was an even greater number of car—257 per 1,000 persons—than previous reports of 195 cars per 1,000 persons in 2014 (City of Copenhagen, 2017b). The national data implies a 31 percent increase in car ownership inside Copenhagen between 2014 and 2018, or 11,775 new cars added to the city per year. This rate of car ownership notwithstanding, the increased number of cars threatens the city's green mobility metrics.

Copenhagen's inner ring suburbs, which span a 120° crescent on the western edge of Copenhagen, have a rate of car ownership that is significantly higher than that of Copenhagen, at 449 cars per 1,000 persons (Statistics Denmark, 2018a). The inner ring suburbs, which enjoy excellent regional rail connections and a decent bus service, are also poised for shifting more trips to the cycle superhighways, but cycling rates have plateaued at less than 20 percent since 2012 and most of these are short school run trips (Capital Region of Denmark, 2017). The distance from Copenhagen matters as well. Further out from the core, cycling rates in the suburbs drop to 10 percent. While these are relatively high rates of cycling in relation to most of Europe and North America, compared to Copenhagen these cycling rates are lower and the Capital Region of Denmark (ibid.) warns that "figures are stagnant."

The commuter shed for Copenhagen extends to all of the Island of Zealand, which, with a population of 2.65 million, is highly integrated into the Copenhagen regional economy (Statistics Denmark, 2018b; Winther and Hansen, 2006). Car ownership rates in Western Zealand, where there are more working- and middle-class suburbs and also the older market towns of Roskilde and Koge, is 429 cars per 1,000 persons. This increases to 466 cars per 1,000 further outward in southern Zealand.

So far all of these car ownership rates hover closely to the national car ownership level of 438 cars per 1,000 persons, but are significantly higher than car ownership in Copenhagen (Statistics Denmark, 2018a). However, in the wealthier and more politically conservative northern tier of Copenhagen's suburbs car ownership rises to 520 cars per 1,000 persons, the highest in any part of Denmark and resembling Germany's car ownership rate, and is more than double the rate for the municipality of Copenhagen (ibid.). In certain suburbs, rates of car ownership even approach 600–700 cars per 1,000 persons, resembling car ownership rates in US metropolitan areas.

Throughout Denmark car ownership is increasing. Some 220,00 cars were sold in 2017, a record year, and between 2008 and 2018 the growth rate was 20 percent (Danish Road Directorate, 2017; Astrup, 2018a). According to Statistics Denmark, since the minimum age requirement for a driving license was dropped to seventeen younger Danes are buying cars as never before (ibid.). There are now five times as many cars in Denmark as there were in 1962 (Politiken, 2016) when the proliferation of cars began.

Denmark is not the only country in Europe today where car ownership and use are perceptibly increasing, but it is one of the fastest growing (Eurostat, 2018). While in some peer countries car ownership may be slowing or has reached saturation point, in Denmark the combination of increased household incomes and the relatively lower cost of owning and operating cars might be pushing up car ownership, and this trend is no different from historic increases in car ownership in the United States and Europe, or the emerging global middle class in Asia or Latin America (Schipper, 2010; Martin, 2015). Since 2012, at the tail end of the Great Recession, driving costs shrank throughout Denmark, and driving increased by 10 percent (Danish Road Directorate, 2017).

Relatively lower fuel prices, reduced car taxes (reduced from 180 percent to 150 percent of the cost of a car in 2015), and the subsidy for longer distance commuters have combined to increase the popularity of driving (Danish Road Directorate, 2017). From 2012 fuel prices decreased by 20 percent (mainly due to declining global oil prices) and smaller "micro" cars were made cheaper by changes in the Danish car tax that rewarded better fuel economy. Many people were able to afford their first car while others wanted to buy bigger cars because registration taxes on larger cars had also been reduced.

The way in which Danish taxes are structured encourages and subsidizes car ownership and use for commuting if the distance to be travelled from home is greater than 12 kilometers. This subsidy was initially meant to address unemployment in Denmark, especially in rural areas, by encouraging people to look further away from their home for work. One perhaps unintended effect of car ownership is that more car commuters are finding it less expensive to relocate away from their places of work. Many buy a car to cover the distance. Many might already own a car, with sunk costs in ownership, and have already paid the high car tax. With a weaker land use regulatory framework and neglect of the Finger Plan, jobs are also decentralized and dispersed, and therefore access via rail stations is less convenient. In 2017 the Danish Road Directorate suggested that if a job is located more than 600 meters (0.4 mile) away from home, workers are more likely to drive.

Meanwhile, some Danes, like their peers in the United States, increasingly believe that it is necessary to have a car when they have children (Danish Road Directorate, 2017). In Denmark, 83 percent of families with children have cars, compared to just 53 percent of households without children. Families with children move to the suburbs because bigger houses are cheaper. Car dependency sets in and many households buy a second car.

While most people commuting from the suburbs to the center of Copenhagen do not drive, in the suburbs there is more lateral, suburb-to-suburb commuting, as well as increased reverse commuting from Copenhagen to distant work places. At an average one-way distance of 19 kilometers (almost 12 miles), there is a lot of car commuting in Copenhagen's suburbs, and congestion on the Ring 3 motorway (see Figure 7.1), which crosses all five Fingers of the Finger Plan and bypasses the inner suburbs, is especially chronic.

The normalization of the car is begetting more cars (Danish Road Directorate, 2017). More cars are fitted with comfortable accessories and technology, making driving more attractive despite traffic congestion. As more families with children buy and drive cars, their children are acclimatized into a car culture, and learn to drive when they reach the legal age to do so. Other families feel marginalized and compelled to get a car, as it becomes a cultural norm to have a car for the "good life." In built environments that make the car easy to use, the car becomes further identified with status, identity, freedom, flexibility, and the liberty of individual choice to be entertained or travel at will. To be sure, this good life can be achieved in compact urban areas where there are good public transport and cycling systems, and even the Danish Road Directorate (ibid.) acknowledges this. More cars and more people driving is not inevitable.

Yet public policies and politics in Copenhagen suburbs and throughout Denmark have deliberately made owning cars easier and less expensive than previously. This has in some ways subsidized longer car commutes, avoided petrol tax hikes, and generally has supported road building and parking structures. This may result in further essentialization of the car in Danish politics.

Copenhagen's climate gap

Given that we introduced this book with the rationale that Copenhagen is an important bellwether for climate change and transport policy, we must briefly describe how the plateauing of cycling and the increase in car ownership have made Copenhagen's efforts to reduce transport emissions problematic. To be sure, there is no question that high cycling rates, low car ownership and low car use have helped Copenhagen to achieve some of the lowest per capita car emissions (0.575 metric tons of CO_2 equivalent emissions per capita in 2015) among peer cities around the world (City of Copenhagen, 2016b). Yet car emissions in Copenhagen are rising, and Copenhagen's climate plan does not map a clear path for dealing with this. Something else is rotten.

Briefly, the Copenhagen 2025 Climate Plan, adopted in 2012 and updated in 2016, is recognized globally as one of the most far-reaching city-scale efforts to mitigate greenhouse gas emissions (C40 Cities, 2013; European Commission, 2013; London School of Economics and Political Science, 2014). Some of Copenhagen's most impressive green indicators relate to the 2025 Climate Plan, including shifting electricity and heat production away from coal to renewable biomass and wind energy. Policies underwriting residential building upgrades and renovations, new building codes, and greening municipal buildings and vehicles are other conservation measures that have attracted global praise and are also part of Copenhagen's green branding (City of Copenhagen and State of Green, 2014).

Yet there is little synchronicity between the goals of Copenhagen's climate plan and the politics of mobility in Copenhagen. Emissions from transport, which were 34 percent of all of Copenhagen's locally produced emissions in 2010, have stayed constant, and car emissions, which are 23–24 percent of Copenhagen's locally produced emissions, have nudged upward (City of Copenhagen, 2016b). The 2025 Climate Plan called for reducing the mode share of cars from 33 percent to 25 percent by 2015 (ibid., 2012). Green mobility modes—cycling, transit, and walking—were to make up 75 percent of all trips.

Despite these laudable goals, as of 2018 the mobility target was elusive. The share for cars was up a notch to 34 percent of all trips with no immediate signs of decreasing. Copenhagen's green mobility goals for climate were hitched to the toll ring (City of Copenhagen, 2012, 2016). But the climate plan politely referred to "uncertainties" about the toll ring and the linkage between the toll ring and green mobility goals was always loose at best. As we outline in Chapter 5, the toll ring was scuttled, and the goals were then hitched to parking reform. As we describe in Chapter 6, parking reform stalled. Copenhagen then moved the goalpost, and the 25 percent goal for car trips was extended to 2025.

Copenhagen's 2025 Climate Plan reflects the outcome of the city's contradictory politics of mobility. The plan includes enthusiastic celebration of cycling and public transit. It acknowledges that Copenhagen's transport emissions would be much higher were it not for such high rates of cycling, and it calls for expanding cycling to 50 percent of all work and education trips, in synchronicity with the city's parallel bicycle plans. Yet the climate plan describes little to nothing about car restraint, the toll ring, or the need for parking reform.

Even with 90 percent of Copenhagen's transport emissions and pollution from cars entering, leaving, or passing through the city, the 2025 Climate Plan does not challenge the car (City of Copenhagen, 2015b). Instead, with resignation, the 2016

Copenhagen Climate Plan Roadmap, an update of the 2025 Climate Plan, observed that "more Danes will own cars in the future" and that household car ownership is increasing in Copenhagen, which will translate into a 20 percent growth rate in cars in the city because of accompanying population growth. The authors of the 2025 Climate Plan were not wrong when they declared transport the city's biggest climate challenge (ibid., 2012, 2016b).

Yet, by clever accounting, Copenhagen's 2025 Climate Plan calculates future car emissions against reduced emissions in other city sectors like electricity and heating. It is relevant to take a moment to consider what Copenhagen's car emissions are being offset against, because this is also disconcerting given the urgency of climate change.

Historically Copenhagen's heat and electricity were produced by burning coal (and before coal, oil). Copenhagen's 2025 Climate Plan, like most climate plans around the world, calls for coal to be phased out. By 2025 coal is supposed to be replaced by a combination of renewable biomass fuels (straw, wood chips) and wind power for electricity. Copenhagen's district heating system, which was a magnificent public works undertaking, will be fueled by "renewable" wood pellets, straw, and household waste. Combustion of this wood and waste will be in specialized "combined heat and power" stations, which will generate heat and electricity simultaneously. Meanwhile, new offshore wind farms in the Baltic (if built) will generate abundant renewable electricity. During windy periods, when more electricity might be generated than Copenhagen demands, the excess will be sold to other parts of Denmark or to Germany.

All things considered, emissions from biomass and household waste are less than coal, and wind is considered zero emissions. But Copenhagen's 2025 Climate Plan is not a decarbonization plan. Instead, it is defined as a "carbon-neutral" plan. According to the City of Copenhagen (2012), carbon neutrality is achieved when consumption of renewable energy like biomass or wind is equal to the total energy consumption within the city, even if some of that energy produces greenhouse gas emissions.

Emissions from private cars can be offset by biomass energy, which is renewable and lower emitting than coal. In turn, the emissions from biomass are offset by high rates of wind energy production, including exporting excess electricity generated during periods of sustained high winds, and ultimately with battery storage of electricity generated by wind (which is proposed but not yet deployed to utility-scale mass consumption).

The risky point here is that Copenhagen's 2025 carbon neutrality scheme uses wind and renewable biomass energy to offset increased carbon emissions from private cars. This means that Copenhagen plans to build excess renewable energy capacity (more biomass, more wind, and at some point utility-scale batteries) than it really needs, and to sell that surplus to offset against increased private car use. Car emissions are masked, and not reduced, in this plan.

Furthermore, the calculation of the offsets needed for car emissions is misleading. While accounting for carbon emissions released by burning wood and waste, the agro-forestry industrial system that supplies the fuels also releases carbon and this is not part of the plan's calculations. These are displaced emissions, and unaccounted for in the city's climate plan. Similarly, wind power, made from concrete, steel, and petroleum has its own set of emissions, and the yet-to-be deployed utility-scale batteries will have their own set of emissions and toxic pollution issues to acknowledge (Zehner, 2012). Furthermore, as with most climate plans, Copenhagen's displaced emissions stemming from foreign car manufacturing, oil extraction, oil refining, and other corollary life-cycle impacts are unaccounted for.

In 2018 there was also another gap in Copenhagen's carbon neutral scheme. The anticipated new wind energy had not yet come on stream, and was delayed by land use and energy politics in Denmark (City Climate Planner, 2016). Few sites are made available for utility-scale wind farms in Denmark without a major political fight over land use. Offshore sites in the Baltic are not controlled by Copenhagen's public utility. There is no guarantee of utility-scale wind production coming on stream in a timely manner. Copenhagen's plan to be "carbon neutral" by 2025 is off track.

Conclusion: Copenhagen at a crossroads

Copenhagen promotes a future of higher rates of cycling and greater car restraint, but has not figured out exactly how to do it. The city is juxtaposed against the suburbs and the rest of Denmark, where car use is increasing, and cycling rates are stagnant. There are new, increasingly affluent demands for cars in the city and cars vie for what might be future bicycle or transit space. Ultimately politics is going to settle these street fights and so we now turn to the second part of the book, which examines Copenhagen's politics of mobility.

References

Astrup, J. 2018a. "Traffic Change: Bikes Are Losing to Cars." *Politiken*, 4 April.

Astrup, J. 2018b. "Will the Traffic in the Bike Lanes Become More Risky? Electric Vehicles Can Be Allowed in the Cycle Paths." *Politiken*, 6 December.

Bennetsen, N. M. and Magelund, J. O. 2018. "Planning for Sustainable Mobilities: Creating New Futures or Doing What Is Possible?" In M. Freudendal-Pedersen, K. Hartmann-Petersen, and E. L. P. Fjalland, eds., *Experiencing Networked Urban Mobilities*. New York, Routledge.

C40 Cities. 2013. *Award Profiles: Copenhagen—CPH Climate Plan 2025*. New York, C40 Cities. Available at www.c40.org/awards/2013-awards/profiles/3 (accessed 30 June 2018).

Carstensen, T. A. and Ebert, A.-K. 2012. "Cycling Cultures in Northern Europe: From 'Golden Age' to 'Renaissance.'" In J. Parkin, ed., *Transport and Sustainability*, vol. 1: *Cycling and Sustainability*. Bradford, Emerald Group.

Capital Region of Denmark (Region Hovedstaden). 2017. *Cycling Report for the Capital Region*. Hillerød, Centre for Regional Development.

City of Copenhagen, 2011. *Good, Better, Best: The City of Copenhagen's Bicycle Strategy 2011–2025*. Copenhagen, Copenhagen Technical and Environmental Administration.

City of Copenhagen. 2015a. *Copenhagen City of Cyclists: The Bicycle Count 2014*. Copenhagen, Copenhagen Technical and Environmental Administration.

City of Copenhagen. 2015b. *Copenhagen Green Accounts 2014*. Copenhagen, Technical and Environmental Administration.

City of Copenhagen. 2016a. *Status of Copenhagen 2016: Key Figures*. Copenhagen, City of Copenhagen.

City of Copenhagen. 2016b. *CPH 2025 Climate Plan: Roadmap 2017–2020*. Copenhagen, Technical and Environmental Administration.

City of Copenhagen. 2017a. *Copenhagen City of Cyclists: The Bicycle Count 2016*. Copenhagen, Copenhagen Technical and Environmental Administration.

City of Copenhagen. 2017b. *Traffic in Copenhagen: Traffic Figures 2010–2014*. Copenhagen, Technical and Environmental Administration.

City of Copenhagen. 2017c. *Cycle Track Priority Plan (2017–2025)*. Copenhagen, Copenhagen Technical and Environmental Administration.

City of Copenhagen and State of Green, 2014. *CPH 2025 Climate Plan: Solutions for Sustainable Cities*. Copenhagen, City of Copenhagen and State of Green.

Colville-Andersen, M. 2014. *The E-bike Sceptic*. Copenhagen, Copenhagenize Design Company. Available at www.copenhagenize.com/2014/02/the-e-bike-sceptic.html (accessed 7 June 2018).

Colville-Andersen, M. 2015. Presentation by Mikeal Colville-Andersen to Copenhagenize Master Class, 23, 24, 25 June, Copenhagen, Copenhagenize Design Company.

Colville-Andersen, M. 2018. *Copenhagenize: The Definitive Guide to Global Bicycle Urbanism*. Washington, DC, Island Press.

Danish Parliament. 2016. *Future of Cycling Policy Summit (Fremtidens Cykelpolitik)*, 29 August. Copenhagen, Danish Parliament.

Danish Road Directorate. 2017. *Driving Forces: Why Is Road Traffic Growing in Denmark?* Copenhagen, Danish Road Directorate.

European Commission. 2013. *Copenhagen: European Green Capital 2014*. Luxemburg: European Commission.

Eurostats. 2018. *Passenger Cars in the EU*. Available at https://ec.europa.eu/eurostat/statistics-explained/index.php?title=Passenger_cars_in_the_EU (accessed 13 June 2018).

Freudendal-Pedersen, M. 2015. "Whose Commons are Mobilities Spaces? The Case of Copenhagen's Cyclists." *ACME: An International E-Journal for Critical Geographies*, 14: 598–621.

Gössling, S. 2013. "Urban Transport Transitions: Copenhagen, City of Cyclists." *Journal of Transport Geography*, 33: 196–206.

Kabell, M. 2015. Cycling in Copenhagen: Presentation to Copenhagenize Master Class, June. Copenhagen, Copenhagenize Design Company.

Larsen, J. 2016. "The Making of a Pro-cycling City: Social Practices and Bicycle Mobilities." *Environment and Planning A*, 49(4): 876–892.

Larsen, J. and Funk, O. 2018. "Inhabiting Infrastructures: The Case of Cycling in Copenhagen." In M. Freudendale-Pedersen, K. Hartmann-Petersen, and E. L. P. Fjalland, eds., *Experiencing Networked Urban Mobilities*. New York, Routledge.

London School of Economics and Political Science (LSE). 2014. *Copenhagen, Green Economy Leader*. London, LSE.

Martin, G. 2015. "Global Automobility and Social Ecological Sustainability." In Alan Walks, ed., *The Urban Political Economy and Ecology of Automobility: Driving Cities, Driving Inequality, Driving Politics*. New York, Routledge.

Movia. 2016. *Trafikplan 2016 (Traffic Plan 2016)*. Copenhagen, Movia.

Oldenziel, R., Emanuel, M., de la Bruheze, A. A., and Veraart, F., eds. 2016. *Cycling Cities: The European Experience*. Eindhoven, Foundation for the History of Technology.

Politiken. 2016. "There Are 5x More Cars in Denmark Today than in 1962." *Politiken*, 18 March.

Politiken. 2018. "Minister Fails Expert Warnings: Fast Electric Bikes Go into the Bike Paths." *Politiken*, 26 June.

Pucher, J. and Buehler, Ralph 2012. *City Cycling*. Cambridge, MA, MIT Press.

Safe Routes to School. N/D. *The Decline of Walking and Bicycling*. Washington, DC, Pedestrian and Bicycle Information Center. Available at http://guide.saferoutesinfo.org/introduction/the_decline_of_walking_and_bicycling.cfm (accessed 5 June 2018).

Schipper, L. 2010 "Car Crazy: The Perils of Asia's Hyper-Motorization." Global Asia,*Journal of the Asia Foundation* 4. Available at www.globalasia.org/l.php?c=e243 (accessed 8 March 2011).

Statistics Denmark. 2018a. *Car Ownership*. Available at www.statistikbanken.dk/10220 (accessed 5 June 2018).

Statistics Denmark. 2018b. *Population and Elections*. Copenhagen, Statistics Denmark.

Tumlin, J. 2012. *Sustainable Transportation Planning: Tools for Creating Vibrant, Healthy, and Resilient Communities*. Hoboken, John Wiley and Sons.

Vedel, S. E., Jacobsen, J. B., and Skov-Petersen, H. 2017. "Bicyclists' Preferences for Route Characteristics and Crowding in Copenhagen: A Choice Experiment Study of Commuters." *Transportation Research Part A: Policy and Practice*, 100: 53–64.

Winther, L. and Hansen, H. K. 2006. "The Economic Geographies of the Outer City: Industrial Dynamics and Imaginary Spaces of Location in Copenhagen." *European Planning Studies*, 14: 1387–1406.

Zehner, O. 2012. *Green Illusions: The Dirty Secret of Clean Energy and the Future of Environmentalism*. Lincoln, University of Nebraska Press.

Interviews held by Jason Henderson from 2015 to 2017 with transportation experts in Copenhagen

City Climate Planner (1), 2016.
City Transport Planner (1), 2016.
City Transport Planners (2 and 3), 2016.
City Transport Planner (4), 2016.
Regional Planner (1), 2016.
Transport Consultant (2), 2016.
Transport Scholar (2) 2016.

4

THE POLITICS OF MOBILITY IN COPENHAGEN

Introduction

Copenhagen has an impeccable bicycle system and high rates of cycling, and images of cycle tracks are used in promoting the city. Notably absent from Copenhagen's promotional and branding material are images of the car. Yet the car looms large in Copenhagen's politics of mobility. Car ownership is rising, and if cycling is to expand it means that space occupied by the car must be removed. Today flashpoints in Copenhagen include how street space is physically allocated, debates about pricing and metering of the car, the car parking spaces, and broader debates about how bicycles, cars, pedestrians, and public transit should circulate in the city, by how much, and exactly where. While variegated and shaped by local place-based conditions, we argue that three broad ideologies of mobility—Left/Progressive, Neoliberal, and Right/Conservative—are shaping these debates and their outcomes (see Table 4.1).

In this chapter, we examine this three-way politics of mobility. This framework situates Chapters 5–7, in which we analyze Copenhagen's congestion pricing debate, debates about car parking on the streets and in new developments, and emerging flashpoints in the politics of mobility surrounding the proposed harbor tunnel on Copenhagen's east side, the opening of the Metro City Ring, and proposals to remove cars from parts of the city center.

Left/Progressive politics of mobility

Copenhagen is the core of Left/Progressive politics in Denmark as well as the core of cycling. By Left/Progressive we denote a politics that is willing to challenge the system of capitalism (hence Left) while articulating a strong belief in the ability of government and the public sector to work meaningfully for people (hence

TABLE 4.1 Copenhagen's politics of mobility matrix (2018)

	Political party	Mobility slogan	Mobility vision	Role of government
Left/ Progressive	Red–Green Alliance	Car-free city life	Right to the green mobility city	Government promotion of bicycle space and car restraint
	Socialist People's Party	Leave the car at home	Right to the green mobility city	Government promotion of bicycle space and car restraint
	The Alternative	2020–2030 is last chance on global warming	Right to the green mobility city; new technology emphasis	Government promotion of bicycle space and car restraint
Neoliberal	Social Liberal Party	Denmark should be the global leader in cycling	Neoliberal with progressive mobility framing; creative class party	Government promotion of bicycle space and car restraint
	Social Democrats	Greener and safer traffic in Copenhagen	Mobility for capital accumulation and land development; cars are part of economic growth; families need cars	Government promotion of bicycle space, but actively accommodate cars with parking/ new harbor tunnel
	Liberals	A city with economic growth and the possibility to own a car	Mobility for capital accumulation and land development; cars are part of economic growth; families need cars	Government promotion of some bicycle space, but accommodate cars with parking/new harbor tunnel
	Liberal Alliance	Cheaper cars increase safety for families	Mobility for capital accumulation and land development; cars are part of economic growth; families need cars	No more car restraint; eliminate car tax; government should build more car parking and build new harbor tunnel
Right/ Conservative	Danish People's Party	Freedom of choice	Private cars are necessary; Bicycle space should not impede cars	No more car restraint; government should build more car parking and build new harbor tunnel with no tax increase
	Conservative People's Party	Free to park in the city	Private cars are necessary; bicycle space should not impede cars	Cycling is acceptable but no more car restraint; government should build more car parking and build new harbor tunnel

Progressive) (Esping-Andersen, 1990; Harvey, 2012; Emmenegger *et al.*, 2015; Keman, 2017). The political will to challenge capitalism can be found all around the globe, but is variegated with different pronunciations and inflection points (Harvey, 1996). We emphasize Left in our analysis because of a tendency, especially in the United States, to use the term Progressive as a catch-all that obfuscates differences between left-leaning and neoliberal politics which might support cycling and public transit. For example, in the United States the political left and the neoliberals might both articulate a progressive mobility agenda of walkability, cycling, and public transit, but disagree about the extent of private sector involvement, labor politics, and pricing (Henderson, 2013). The Left/Progressives would endorse less private sector control, more leveling of wages and work conditions between labor and management, and redistributive forms of pricing.

It is especially notable that in many cities around the world similar debates about cycling and the car are taking place, permeated by strong currents of Left/Progressive politics (Brenner, *et al.*, 2012; Blanco *et al.*, 2018; Verlinghieri and Venturini, 2018; Sosa López and Montero, 2018).

In Copenhagen's politics of mobility, a key distinction between the Left/Progressives and the Neoliberals (as well as the Right/Conservatives) is that while there is almost unanimous appreciation for the bicycle, the Left/Progressives seek much greater car restraint. The Neoliberals and the Right/Conservatives balk at greater car restraint even if it brings about more cycling, and this means that the political ideologies contain very different visions of mobility in Copenhagen. Muddying this distinction somewhat, the Social Liberals advocate for expanding cycling and car restraint, but depart from the Left/Progressives by avoiding a left-leaning critique of capitalism (Radikale Venstre, 2018).

Copenhagen's Left/Progressive political parties include the Red-Green Alliance, the Socialist People's Party, and the Alternative, all of which advocate for expanding Copenhagen's bicycle system while also implementing car restraint. Together, these three parties controlled twenty-two (40 percent) of the fifty-five seats in Copenhagen's city council in 2018 (City of Copenhagen, 2018). The Red-Green Alliance is the largest of the three parties, and garnered a 20 percent share of the vote in the 2017 municipal elections. The Red-Green Alliance is also the second-largest political party in Copenhagen after the Social Democrats (who took fifteen seats, or a 27 percent share of the vote).

Since 2013, and because the party has a solid second-place footing in Copenhagen's electoral process, the Red-Green Alliance has steered Copenhagen's Technical and Environmental Committee, the municipal agency that plans for cycling, car traffic, and how street space is allocated (City of Copenhagen, 2018). With its slogan "car-free city life," the Red-Green Alliance is recognized by Copenhagen city planners and advocates as a forthright promoter of both cycling and car restraint (City Transport Planner (1), 2016; Enhedslisten, 2018a). In addition to green mobility, the Red-Green Alliance also promotes the traditional leftist tenants of social democracy such as universal social welfare. The party openly challenges capitalism, and identifies in opposition to neoliberalism (ibid., 2018b).

The Socialist People's Party, an older spin-off from the former Communist Party, also challenges capitalism from the left, but is not recognized as being particularly visionary about cycling (City Transport Planner (1), 2016). The party does suggest that people "leave the car at home," and generally backs the Red-Green Alliance on green mobility politics (Socialistisk Folkeparti, 2018). The Alternative, the third and smallest partner in Copenhagen's municipal Left/Progressive coalition, does not articulate a blunt leftist critique of capitalism like the Red-Green Alliance and the Socialist People's Party, yet does bend towards critiques of capitalist growth policies when urgently warning that the decade 2020–2030 will be our final chance to take action against global warming (Alternative Copenhagen, 2017). The Alternative promotes the expansion of cycling and more far-reaching car restraint, but also expresses hope in new technologies such as shared mobility with electric driverless cars.

The Social Liberals, also a small party but not leftist, does generally align with the Left/Progressives on green mobility, but deviates from the Left/Progressives on development issues, with a strong green growth profile (Radikale Venstre, 2017). Defying easy categorization, this party might be classified as a hybrid Neoliberal and Progressive party.

It is worth considering that if the Social Democrats were aligned with the other Left/Progressive parties, the Left/Progressives would have a super-majority in Copenhagen's city council, with 67 percent of the seats. Yet as we have already described in previous chapters the Social Democrats have been ambivalent on the subject of cycling and the car and have most recently drifted towards neoliberal policies. In 2018 this meant that the Left/Progressives were unable to establish a governing coalition that would put a Left/Progressive incumbent in the Lord Mayor's chair. Based on their perspective that Denmark should be a world leader in cycling, and their support for car restraint, the Social Liberals might at times be a reliable ally for the Left/Progressives, but with only five seats on the city council, this still falls short of the threshold needed to create a governing coalition.

Instead, the Social Democrats, with support from a handful of smaller Neoliberal and Right/Conservative parties, controls the Lord Mayor and the city's Finance Department, which oversees land development, the Metro, and the city budget (impacting funding for cycle track expansion and other green mobility policies). However, due to the Danish practice of consensus and compromise, the Left/Progressives have not been absent from planning.

The Left/Progressives, summoning a social democratic heritage, promote an active government that not only seeks to expand cycling but that funds and improves public transit, makes Copenhagen housing more affordable by intervening in the market, and physically limits the number of cars permitted in the city. Progressive threads stem from the position that the government should invest in mobility policies that ensure environmental sustainability, social equity, and social welfare, while limiting the privatization of public services and public spaces (the Red-Green Alliance and the Socialist People's Party object to the privatization of transit and other public assets). Sometimes invoking modernist social democratic values, the Left/Progressives believe that urban planning can be used to achieve social goals, is necessary for an orderly response to global warming, and that new housing and urban development should not include financial speculation or be controlled by corporations (Enhedslisten, 2018c)

The Red-Green Alliance, the Socialist People's Party, and the Alternative are particularly critical about neoliberal "green growth" policies such as using cycling as a brand for Copenhagen to attract tourism and encourage knowledge-based companies to locate in the city and region (Enhedslisten, 2018c). They are concerned that neoliberal green growth is neither truly green, nor equitable. As affluent people seek out urban living near good cycling routes and public transit, the areas that are primed for cycling and new public transit schemes increase in market value, especially because of deregulation of housing in Copenhagen (Hansen *et al.*, 2001; Larsen and Lund Hansen, 2008; Andersen and Winther, 2010; Baeten *et al.*, 2015). Evictions and rent escalation in the housing market can push the working poor and middle class out of the city.

While lower-income Copenhageners might be displaced, many affluent Copenhageners bring cars with them when moving into the city, or buy cars, and demand more expensive housing with adjacent car parking (City of Copenhagen, 2016). This leads to more cars, demand for more space for cars, less space for housing, and adds to the overall cost of housing (Shoup, 2005). For the Left/Progressives, if cycling and green mobility elevate housing costs as part of the package of urban livability then the city must intervene not by forfeiting green mobility but by preserving the existing affordable housing stock and requiring more social housing be built in the city, preferably without parking for cars.

Thus, Copenhagen's Left/Progressives align with broader notions of the "right to the city" whereby non-property-owning classes have the right to remain in cities that are becoming increasingly exclusive due to gentrification and displacement (Harvey, 2012; Brenner *et al.*, 2012; Beitel, 2013). Essentially this translates into supporting affordable housing policies and social welfare programs that keep a range of different categories of people—unemployed or underemployed, artists, university students, lower-class service workers, renters, pensioners, and aspirational middle-class families seeking additional space in which to bring up their children—in the city and also include these people in decisions about the city.

Inclusivity and the "right to the city" are critically important for how Copenhagen's Left/Progressives think about green mobility and the long-term viability of progressive urban politics (Larsen and Lund Hansen, 2012). Without a strong housing and social services program keeping residents in the city, many displaced residents find less expensive housing far from the urban core, in the suburbs and satellite towns with less access to jobs and nearby services. Since the working class cannot afford to live in the vibrant and chic capital city, and are often pushed into car-dependent configurations in the suburbs, resentment towards higher-cost Copenhagen can eventually undermine green mobility goals.

We discuss these feelings of resentment in greater detail later in the chapter, but it is important to consider the dynamic between Left/Progressive mobility and the antagonism that might stem from otherwise left-of-center inhabitants. Displaced former Copenhageners may come to resent the encroachment of cycling or public transit in the car spaces which they now might use, and new opposition arises against car restraint policies such as congestion pricing, parking limits, or raising taxes and fees on cars (examples of which appear throughout this book). Instead of progressive public policy addressing global warming and social equity, green mobility and car restraint policies are flipped around as regressive taxation. Many displaced former Copenhageners might then realign with more Right/Conservative parties such as the Danish People's Party, which exhibits right-wing populist tendencies that are critical of green mobility, and appeals to some working-class and lower-income voters who own cars (Jyllands-Posten, 2016).

In other parts of Europe this politics of resentment could be seen in the "movement des gilets jaunes" ("yellow vest") protests that were ongoing in Paris in 2019, and which were sparked by populist anger over an increase in the French petrol tax (Kimmelman, 2018; Watts, 2018). The backlash touched on a populist critique of the Parisian urban elite and the neoliberal hue of green mobility in Paris. The protestors, many of whom identified with Right/Conservative discourses, were fomented by outrage over the cost of living, stagnation of wages, and declining quality of life, while in Paris the elite seemed to get on well with public transit and new cycling policies. As in France, this politics of resentment can be found in certain suburban areas in Copenhagen and indeed in other parts of Denmark (Lidegaard, 2009). It spans from anti-immigration politics and debates about Danish culture to resentment over car taxes, and stokes right-wing populist outrage at the municipality of Copenhagen and the cycling culture that it promotes.

Copenhagen's Left/Progressive response to the green mobility-housing imbalance and to the potential right-wing backlash is to assert that the spatial range of cycling must be inclusive, meaning that a fully rounded, robust Left/Progressive green mobility includes affordable housing and social services that are universally accessible within the existing city as well as in the suburbs (Larsen and Lund Hansen, 2012; Andersen et al., 2012; Enhedslisten, 2018b). Ultimately, the spirit of Left/Progressive mobility politics is that it should be possible to cycle everywhere, be that in the city, suburb, town, village, or countryside. With public and affordable passenger rail as armature, the cycling city can be anywhere. At the same time there must be an easy and affordable choice to live car-free anywhere, but especially within Copenhagen. For Copenhagen's Left/Progressive politics of mobility the right to the city is also the right to a green mobility city.

Neoliberal mobility in Copenhagen

As we saw in Chapter 3, branding Copenhagen meant that the neoliberal politics of mobility invokes progressive concerns over mobility, global warming and urban livability, but not a leftist critique of capitalism, nor of the car. Just as social democracy is antithetical to neoliberalism, Copenhagen's Neoliberals—comprising the Liberals, the Social Democrats, and the Liberal Alliance—do not support the

Left/Progressive "right to the city" discourse, and instead emphasize market-oriented housing and mobility (Anders, 2017; Venstre, 2018; Liberal Alliance, 2018a). For the Neoliberals, the car is part of economic growth in Copenhagen. A new, expensive car suggests success and is a favorable item of conspicuous consumption, and Neoliberal voters (especially Liberal Alliance voters, but also the Liberals) are competitive and ambitious, defying Danish social democracy and the core value of humbleness and aversion to personal arrogance (ibid., 2018b; Venstre, 2018). For the Liberal party and the Liberal Alliance, the political project is ultimately to dismantle Danish social democracy (Liberal Alliance, 2018b). For the Social Democrats as well as the Social Liberals, neoliberal mobility is not as far-reaching; instead, it is part of a political project to use the market to underwrite social welfare. What binds all of these parties together is their adherence to the green growth position and their reluctance to implement any further car restraint in Copenhagen (with the exception of the elusive Social Liberals).

The Liberals, the junior partner in the neoliberal political circuitry of Copenhagen, have a confusing name because "Venstre" means "left." It is best defined as the Liberal Party of Denmark (the party's origins date back to the nineteenth century when liberal economic viewpoints were politically to the left of the Conservatives and the Danish monarchy; see Lidegaard, 2009). The Liberals explicitly promote the notion that a city enjoying economic growth and development includes the possibility to own a car (Venstre Copenhagen, 2017). As the traditional Liberal party, the Liberals prefer market-based solutions, privatization, lower taxes, and deregulation of the welfare state—including housing and mobility. Investment in mobility should reflect capital accumulation. Public transit such as the Metro, which brings economic growth to the harbor districts, is favorable, as are government subsidies for electric cars, because the use of more electric cars will stimulate green economic growth (and profit). Underwriting bus routes to lower-income areas, or shifting street space to prioritize bicycles and banning cars may not stimulate economic growth in ways in which the Neoliberals deem fit.

More broadly, in the past two decades the Liberals have deliberately attempted to crack the Danish social welfare model by encouraging wealthier people to send their children to private schools, or to take out private healthcare; the party has also promised to introduce tax cuts. As a result, the party has created a segment of the population unwilling to uphold Danish social democracy (Kuttner, 2008). In terms of mobility, this has involved the semi-privatization of the regional bus system and a reluctance to upgrade the regional and national rail systems. The Liberal Alliance, to the libertarian right of the Liberals in respect of economics, sees the Liberal party as not acting boldly or quickly enough to dismantle social welfare. The Liberal Alliance also envisions that new private technologies such as driverless cars will replace public transit (Danish Ministry of Transport, 2018).

The Social Democrats, whose namesake we have already shown refers to more leftist politics, adds to the confusing labels about neoliberal mobility. The historically left-leaning Social Democratic Party should not conjure up images of neoliberalism; rather, Copenhagen's Social Democrats promote neoliberal economic and urban development polices such as housing deregulation in ways that undercut social democracy (Hansen *et al.*, 2001; Larsen and Lund Hansen, 2008). The Social Democrats consider economic growth as necessary for raising tax revenue to underwrite generous universal social welfare schemes, and therefore they take on entrepreneurial land development schemes and private-public approaches that benefit corporations.

On mobility, the Social Democrats make a vague call for "greener and safer traffic in Copenhagen" but oppose further car restraint because of the belief that the car is necessary for economic growth (Social Democrats in Copenhagen, 2017). The party's historical ambivalence towards cycling and the car, described in Chapter 2, remained part of the Social Democrats' politics of mobility in 2018.

Consistent with the green growth discourse, the Social Democrats and the Liberals seize on the bicycle as an opportunity for urban economic development, and celebrate cycling as a necessary part of a premium livable arrangement connected to a globalized high-speed transport and communications network. Cycling is good for the climate, but neoliberal livability also includes a commodification of walkable, bikeable, and transit-accessible neighborhoods (Harvey, 2008; Henderson, 2013). Yet some of these commodified green mobility districts—such as Copenhagens' harbor—have also become accessible by car, with new housing providing parking for affluent car-owning residents. The municipal Social Democrats promise ample bicycle infrastructure and green mobility, but also promise to preserve street parking, provide less expensive parking, and endorse new private off-street parking in Copenhagen.

Joining the Social Democrats and the Liberals, the more radically neoliberal Liberal Alliance exerts considerable influence on the politics of mobility in Copenhagen. The Liberal Alliance is unabashedly pro-car, arguing that "cheaper cars increase safety and environmental quality (for families)." Significantly, high-ranking members of the Liberal Alliance headed the National Ministry of Transport starting in 2015, as part of the political deal with the Liberals to have a broad Neoliberal-Right/Conservative coalition governing Denmark. The Liberal Alliance taps into a libertarian ethos that questions the role of government and government authority to regulate cars, seeks to eliminate the car tax, and supports pricing of cars only when pricing benefits car users, not as a redistributive tax (City Transport Executive (1), 2016). Furthermore, the Neoliberals, especially the Liberal Alliance, are embracing "tech mobility" such as future electric driverless cars and privatized transportation network services like Uber as a green growth strategy (at early 2019 Uber was still illegal in Denmark, but the Liberal Alliance would like to change that) (see Henderson (2018) for an overview of tech mobility).

Compared to the Social Democrats, the Liberals and the Liberal Alliance are markedly more "revanchist," i.e., they seek to reclaim from Copenhagen's social democratic palimpsest a city for the private market and private property, rather than notions of collective, public, egalitarian livability. This revanchist livability includes discontinuing local government programs that support the urban underclass, using benign neglect by defunding social programs, tax cuts, and supply-side economics, deregulating housing, and using zoning to edge out the working class and unruly classes (such as the unemployed, drug dealers, the homeless, or the mentally ill) from the city center (Smith, 1996).

To be sure, Copenhagen's Social Democrats have also demonstrated neoliberal revanchism. For example, under the social democratic Lord Mayor Jens Kramer (1989–2004), the municipality of Copenhagen pushed out some socially disadvantaged city citizens through neoliberal urban renewal schemes that entailed selling off large stocks of municipally owned rental housing, thereby driving up the price of rental housing in the city by reducing the overall stock (Larsen and Andersen, 2015). This complicates the seemingly leftist profile of the Social Democrats and shows how the party has increasingly supported neoliberal policies in order to satisfy the needs of the increasingly affluent Danish middle class (ibid., 2008; Baeten et al., 2015).

Arguably a goal of the Liberals and Liberal Alliance is to dismantle and privatize social welfare where possible, and this ultimately counters Social Democratic values (Baeten et al., 2015). For different reasons, the Social Democrats deploy neoliberal policies in the hope of supporting the social welfare system; thus, dismantling social welfare is not a Social Democratic Party aspiration. Yet by bargaining on neoliberal terms, Social Democrats navigate perilously. The question is at what point should the party back away from neoliberal policies and shift back to its leftist and progressive roots? We will address that question in the Conclusion to this book after examining how the Social Democrats have maneuvered the flashpoints in Copenhagen's politics of mobility, such as congestion pricing, parking, and the harbor tunnel. Certainly, if they were to reform their politics of mobility, or if enough of their core constituency shifted to Left/Progressives parties, Copenhagen would assuredly tilt to the left, resulting in an expansion of cycling, more car restraint and the promise of a right to the green mobility city.

Right/Conservative politics of mobility

The multi-party system in Denmark, including within Copenhagen's City Hall, meant that following elections in 2018 Copenhagen's neoliberal grouping, with twenty-two seats divided among the Social Democrats, the Liberals, and the Liberal Alliance, were forced to include other parties within the coalition in order to govern. At 40 percent of City Hall seats, the neoliberal balance was strikingly equal to the Left/Progressive proportion in City Hall (this did not include the Social Liberals, who on the basis of politics of mobility, aligned with the Left/Progressives) (City of Copenhagen, 2018). The Neoliberals tipped control of the City Hall in their favor by aligning with the Conservative People's Party and the Danish People's Party, which with six additional votes, pulled the combined Neoliberal-Right/Conservative coalition to a 51 percent majority in the City Hall in 2018. A similar situation could be seen in the national parliament (the Folketing), whereby the Danish People's Party, the second largest party in Denmark, relinquished a leadership role by deferring to the third-placed Liberals to head the state (the Social

Democrats pulled out front, barely, but were unable to form a government because at the national level they formed part of the left-wing faction). For reasons beyond the scope of this book and beyond politics of mobility, in 2018 Denmark was governed, with a slim margin, by a Neoliberal-Right/Conservative coalition.

In Denmark's national politics of mobility, the cocktail of the Liberals, the Danish People's Party, the Liberal Alliance, and the traditional Conservatives, put Copenhagen under the influence of a disproportionately influential Right/Conservative politics. Due to technicalities in the law, the municipalities have limited power over taxation rates, specifically toll and parking charges, which are powers conferred on cities only when enabled by the Danish parliament. Danish law requires that any new tax or fee levied by the regional authorities must first be enabled by the national government (Danish Ministry for Economic Affairs and the Interior, 2018). This is similar to the powers granted to municipalities within the 50 states in the United States, which are subject to enabling legislation by their respective states (for example many US cities would need state enabling legislation to implement tolls or raise parking rates). We will come across instances of parliamentary interference in Copenhagen's politics of mobility as we examine congestion pricing, car parking policy, and the harbor tunnel. In the meantime, though, it is important to get a sense of the ideological underpinnings of the Right/Conservatives and how their ideology appears in Copenhagen.

Traditionally the Conservative People's Party has been part of Danish politics almost as long as the Social Democrats and the Liberals (Lidegaard, 2009). Today their influence has been siphoned off by other parties on the right, especially the Danish People's Party and the Liberal Alliance (Kosiara-Pedersen, 2012). Due to the Social Democrats also pulling to the right, some of their former constituents have aligned with that party as well (Rasmussen, 2014). There is still a conservative ideological presence in national politics, and the Conservatives remain a large presence in Frederiksberg, the high-density enclave surround by Copenhagen, which also has a high cycling mode share (Winther, 2017).

Conservatives in Denmark resemble traditional conservatives in the United States, and place a strong emphasis on cultural values and pride in national identity. While Conservatives are not neoliberalists, there is strong affinity towards the market as the best way to organize economic aspects of society (but not social). Emphasis of conservative politics includes personal responsibility and social responsibility, and this in some ways connects to social democracy as a set of values that includes humbleness and frugality, and taking responsibility for yourself and for others.

Historically, due to conservative values of frugality and humbleness as well as individualism and responsibility for personal mobility, cycling has also been part of Danish conservative culture (Danish Cycling Federation, 2011; Emanuel, 2016) Safety through separating cyclists from traffic, for example, has traits of conservative politics. Even today the Conservative party's mobility platform includes "equal consideration between the forms of traffic," which means that although cars should be accorded space, so too should cyclists (Conservative Party of Copenhagen, 2017a).

There are other similarities between Conservative party goals, Social Democrats' values, and Left/Progressive politics. For example, historically, the Conservatives supported the welfare state and viewed it as a social necessity, but always admonished some of the more collectivist bent of social democracy (Fitzmaurice, 1981). The Conservatives also believe in an active government when it comes to the politics of mobility. When the Conservatives state that a person is "free to park in your city," they are expressing a policy that the government should ensure that there is adequate parking for cars, whether through requiring parking to be made available in zoning for new houses and other developments, or by delineating it on the street next to the curb.

There is one key point in which the Conservatives depart from the Left/Progressives on the politics of mobility: Bicycle space should not be made available at the expense of cars. According to the Conservatives, an active government is necessary to defend car drivers from the encroachment of cyclists and pro-cycling policies. They believe that the government should preserve spaces for driving. The Conservatives claim that private cars are necessary, that the government should accommodate the car, and that a municipality like Copenhagen has no right to restrict driving even if the space is needed for cycling, and the motorist is from outside the city (Conservative Party of Copenhagen, 2017b).

The Danish People's Party, dating back to the 1980s, is not steeped in the old land-owning elite and patrician past of the Conservative Party. Rather, it is a fusion of nationalist identity politics and right-wing populism. The party grew from white underclass resentment in some of Copenhagen's western and southwestern suburbs as well as in Jutland, the mainland Peninsula to the west of Copenhagen and Zealand (Lidegaard, 2009). What separates it from traditional Conservativism is the staunch anti-European Union position, and the broader criticism of globalization and trade. The Danish People's party is not opposed to social welfare—as long as it is for Danes (Sedgwick, 2013). On this point the party thrived in the 2015 elections by siphoning off resentful Social Democratic Party voters who were antagonistic towards their party's internationalist views and openness to providing immigrants with social welfare (Lidegaard, 2009). At one point during the 2015 national elections, the Danish People's Party even ridiculed the Social Democrats for being more to the right on the matter of social welfare cuts (Leonard, 2015). In this instance the Danish People's Party tapped white, working-class Danish resentment by claiming that the Social Democrats prioritized immigrants over the welfare of native Danes.

The Danish People's Party aligns with the Conservatives on matters relating to the car. Invoking "freedom of choice," the Danish People's Party essentializes the car, claiming that private cars are necessary and that cycling spaces should not impede cars. Moreover, on the national scale, the Danish People's Party has promoted national planning that decentralizes and disperses economic growth away from Copenhagen, based on a strong anti-urban, anti-Copenhagen undercurrent, interspersed with anti-immigrant, anti-leftist sentiments, but also suggesting more car-dependent development policies. Interestingly, and departing from rhetoric by the Conservatives, the Liberals, the Social Democrats and the Liberal Alliance, the Danish People's Party suggested that the harbor tunnel, championed by the Neoliberals, would result in too much economic concentration in Copenhagen, and resources might be better aimed elsewhere in Denmark (Hagemeister and Jensen, 2018).

The politics of right-wing mobility and resentment

Together, the Neoliberal-Right/Conservatives have established a loose but pow-erful pro-car coalition in Danish politics, with considerable clout in Copenhagen municipal politics. At the core, this coalition essentializes the car in a variety of ways, and this ultimately undercuts cycling and other forms of green mobility in Copenhagen. Right-wing populist resentment is a fertile area of discourse for empowering this Neoliberal-Right/Conservative coalition and providing it with momentum in the politics of mobility.

Characterizing Copenhagen cyclists as scofflaws resonates with this right-wing resentment. As they queue on Copenhagen streets, cars are constantly overtaken by faster-flowing cyclists because the city was set up to prioritize cyclists. Cyclists might appear to be breaking rules, and sometimes do (because the bicycle is flex-ible and it is easy to bend rules). Rule breaking, mostly "momentumist" is espe-cially prevalent owing to cycle track crowding and bicycle congestion (Colville-Andersen, 2015). The flexibility and minor rule-bending of the bicycle then exacerbates motorists' resentment, which is fertile ground for Right/Conservative populist discourses about cycling and the car (Transport Advocate (1), 2016).

When motorists observe cyclists breaking the rules, so does their resentment towards cyclists increase. Cyclists are perceived as being smug and egotistical, running lights and ignoring the rules (Freudendal-Pedersen, 2015). In an inter-esting twist and exposing a fissure in the Neoliberal-Right/Conservative coali-tion, motorist hostility and resentment is also generated in response to media campaigns that brand and promote cycling as a youthful, hip, environmentally sound, and healthy activity, which suggests that car drivers are none of those things. The cyclists might be healthier than a driver, but they are also said to be arrogant and egotistical. Narratives in the media shift from the neoliberal brand-ing of cycling and promoting cycling infrastructure, to right-wing populist antagonism, that cyclists are scofflaws. There then evolves a discourse of the need to require helmets for safety, and the arrogance of cyclists (see ibid.; Danish Parliament, 2016; Culver, 2018 on helmets).

The Right/Conservatives tap this resentment and this makes expanding car restraint difficult (Freudendal-Pedersen, 2015). The resentment towards Copenha-gens' cyclists can get drawn into a broader anti-urban resentment that is common on the Right/Conservative flank. The Danish People's Party, although increasingly more suburban than rural, frames Danish politics as rural versus urban and chastises the Copenhagen urban elite in its discourse. The party strength is in the swath of exurbia and rural Zealand that is all within the driving commute shed of Copen-hagen (Sedgwick, 2013; Coman, 2015). Forchtner and Kølvraa (2015) describe how the Danish People's Party calls Copenhagen's cycling urbanites "climate demagogues." The Danish Right/Conservative promotion of the car has echoes in the blunt, racialized, anti-urban politics found in many US metropolitan areas (Henderson, 2006, 2013; see also Cramer, 2016).

Outside of electoral and party politics, the Copenhagen division of the Danish police force also deploys a Right/Conservative politics of spite towards cycling (Astrup, 2018). In Denmark, the police are not municipal, but national, with a sub-command for the Capital Region and Zealand. Copenhagen transportation advocates and planners suggest that the political culture of the police is anti-socialist and hostile towards Copenhagen's Left/Progressive urbanites, and therefore to cycling in Copenhagen (City Transport Executive (1), 2016; Transport Consultant (3), 2016; Transport Consultant (2), 2016).

In Denmark, the police can veto proposals for cycle tracks or traffic calming methods in a similar way that fire brigades can oppose traffic safety and bicycle space in the United States (Transport Advocate and Politician, 2016). Cycling bodies in Copenhagen such as the Danish Cycling Federation, and the Red-Green Alliance in particular, advocate for slower speeds, 30 kilometers per hour (18 miles per hour), in the city center and in neighborhoods. Based on research findings they argue that a pedestrian or cyclist is far less likely to die in a collision with a vehicle traveling at 30 kilometers per hour or less.

Yet the decision on speeds lies exclusively in the police realm. According to the Danish police, it is their responsibility and duty to uphold an urban speed limit of no lower than 50 kilometers per hour (31 miles per hour) and that, with few exceptions, lowering speeds poses safety risks. The burden is on cyclists and pedestrians. Cyclists in particular need to obey the rules or they will get hurt.

Furthermore, technical requirements mean that before speeds are reduced, there must be evidence of traffic injury or fatality to pedestrians and cyclists (Transport Consultant, 2015; Transport Scholar (1), 2016). This means that streets onto which cyclists have been discouraged from venturing, but which might be good connecting roads or short cuts, there is little data.

The police respond to Copenhagen's complaints with a shrug. They are directed by the national parliament, since policing is a national not a local responsibility. The national government of 2018, a Neoliberal-Right/Conservative coalition with control of the Ministry of Justice, which oversees police, was slow to change (the police chief for the Copenhagen region is appointed by the Ministry of Justice). Police under a Neoliberal-Right/Conservative governing coalition not only actively promote the car, but perceive Copenhagen's cyclists as undeserving renegades.

Another politics of resentment towards cycling is emerging as a result of parents chauffeuring children by car (Freudendal-Pedersen, 2015). A structural narrative that essentializes the car for families with children is joined to right-wing populist resentment. These narratives include making the car an indispensable tool of parental control over children and their safety since the streets (either in the form of crowded cycle tracks, speeding cars, or buses filled with immigrants) are deemed unsafe. It is "responsible" and "necessary" for parents to drive, echoing neoliberal and conservative narratives of individualism and individual personal responsibility. Next in the structural narrative of chauffeuring children by car, parents emphasize the need to get to school on time, then claim that it is not possible to be part of society without the car. The structural narrative of the need for the car expands and evolves (ibid.). Bad parents let their children cycle alone to school. Good parents drive them to school. Policies that seek to reduce driving, increase parking charges, or reduce car space, are resented.

These structural narratives make it harder to challenge the car politically. The Liberal Alliance has especially harnessed the family car discourse in its efforts to eliminate or reduce the Danish car tax (Liberal Alliance, 2018c). From the Liberal Alliance perspective, the car tax hurts families with children. Families with children, goes the logic, want and need cars, but because of the high car tax, families resentfully purchase smaller, less-safe but cheaper cars.

Tapping into right-wing populism, Danish families should not be forced to the same low quality and low standards of Latin American and Eastern European car owners who are sold similarly small cars. Larger family sedans, as well as higher-standard cars with better mileage, are not affordable to working class families because of the car tax. Danish families with children need heavier and larger cars and SUVs which are "safe" and family-friendly.

The families-need-cars narrative of the Liberal Alliance is effective. In 2015 the Liberal Alliance negotiated in parliament with the larger Liberal and Danish People's parties to reduce the car tax down from 180 percent to 150 percent. In 2018 the Liberal Alliance, deploying the same "families-need-cars" discourse, pushed for eliminating the car tax altogether (Liberal Alliance, 2018c). The party pointed to conveniently agreeable studies, such as an in-house report by the Danish Road Directorate (2017), which stated that owing to increased wealth, more families with children, and social trends, Danes needed to drive and Denmark needed new motorways.

The politics of resentment also spins another way, with resentment towards the politicization of the bicycle as a Left/Progressive icon. For neoliberal branding of Copenhagen's green growth agenda and the iconic cycling city, the preferred discourse is that mobility—including cycling—is non-ideological and non-partisan. People cycle in Copenhagen because it is easy, convenient, and healthy, and not because they want to make a political statement (Carstensen and Ebert, 2012; Consultant (2), 2016). The result is to disassociate cycling with its long history of social democratic and Left/Progressive politics.

If cycling is not political, then it has no political dispute with the car or car policies (Transport Consultant (2), 2016). Depoliticizing cycling detaches cycling from broader critiques of the inequity of capitalism and the urgency of climate change, and away from a direct political critique of the car. Official bicycle literature in Copenhagen avoids dystopian climate change urgency (Gössling, 2013). Copenhagen bicycle plans avoid moral, ethical, and environmental discourses and prefer to invoke the positive image of cycling as a healthy, hip, happy, and fun activity.

This aversion to urgency impacts on car restraint policy because it is also considered impolite to engage in anti-car narratives. Copenhagen's bicycle movement only talks discursively about creating better conditions for bikes, not about taking away car space. Few bicycle publications issued by the City of Copenhagen say much about automobile restraints such as removing car space, removing car parking, or congestion charging and other schemes to reduce driving. The essentializing of the car invariably leads to arguments about depoliticizing cycling and the car, which ultimately severs important political critiques of the car.

Conclusion: now the hard part

Despite resentment and animosity, the existing levels of cycling in Copenhagen are not a political issue within the city itself (Transport Scholar (1), 2016). Cycling is recognized as a fundamental part of Copenhagen and necessary for smooth city operations (and economic development). But expanding the cycling infrastructure is politically charged and, as one bicycle scholar observed, "In Copenhagen the struggle for cycling is far from over" (Emanuel, 2016). Copenhagen's November 2017 Mayor's race indicated that while cycling remained an obvious vote getter, car restraint was not articulated by any strong candidates on the left or right (City Transport Planner (5), 2016). In Copenhagen, it is perfectly acceptable for any ideological stripe—Left/Progressive, Neoliberal, or even Right/Conservative—to extol the good side to cycling, such as frugality and individual propulsion, space efficiency, convenience, and simplicity, with a light nod to the environment. It is not acceptable, however, to suggest taking car space in order to improve or expand cycling. Having done what might be considered the easy part, now Copenhagen must make harder choices. Copenhagen must address bicycle capacity and how to restrain cars, or it will need to expand car capacity, and restrain bicycles (Gössling, 2013; Carstensen et al., 2015). In the next three chapters, we examine the most potent of these hard choices, or "street fights," in Copenhagen's politics of mobility.

References

Alternativet Copenhagen. 2017. *Objectives for Sustainable Development.* Copenhagen, Alternativet.

Anders, L. H. 2017. "Housing Policy Is Major Political and Ideological Battlefield." *Jyllands-Posten*, 16 November.

Andersen, H. T. and Winther, L. 2010. "Crisis in the Resurgent City? The Rise of Copenhagen." *International Journal of Urban and Regional Research*, 34: 693–700.

Andersen, J., Freudendal-Pedersen, M., Koefoed, L., Larsen, J., and Gutzon, H., eds. 2012. *City in Motion: Mobility, Politics, Performity.* Roskilde, Roskilde University Press.

Astrup, S. 2018. "Copenhagen Will Cancel Bike Improvements on Street Traffic." *Politiken*, 13 June.

Baeten, G., Berg, L. D., and Lund Hansen, A. 2015. "Introduction: Neoliberalism and Post-Wefare Nordic States in Transition."*Geografiska Annaler Series B: Human Geography*, 97: 209–212.

Beitel, K. 2013. *Local Protest, Global Movements: Capital, Community, and State in San Francisco.* Philadelphia, Temple University Press.

Berlingske. 2018. "Danish People's Party Will Need to Study Capital Region Project." *B. T.*, 5 October.

Blanco, J., Lucas, K., Schafran, A., Verlinghieri, E., and Apaolaza, R. 2018. "Contested Mobilities in the Latin American Context." *Journal of Transport Geography*, 67: 73–75.

Brenner, N., Marcuse, P., and Mayer, M., eds. 2012. *Cities for People, Not for Profit: Critical Urban Theory and the Right to the City.* Abingdon, Routledge.

Carstensen, T. A. and Ebert, A.-K. 2012. "Cycling Cultures in Northern Europe: From 'Golden Age' to 'Renaissance.'" In J. Parkin, ed., *Transport and Sustainability*, vol. 1: *Cycling and Sustainability.* Bradford, Emerald Group.

Carstensen, T. A., Olafsson, A. S., Bech, N. M., Poulsen, T. S., and Zhao, C. 2015. "The Spatio-Temporal Development of Copenhagen's Bicycle Infrastructure 1912–2013." *Geografisk Tidsskrift—Danish Journal of Geography*, 115: 142–156.

City of Copenhagen. 2016. *Annual Parking Report.* Copenhagen: Technical and Environmental Administration.

City of Copenhagen. 2018. *The City of Copenhagen Government 2018–2021.* Copenhagen, Municipality of Copenhagen.

Colville-Andersen, M. 2015. Presentation by Mikael Colville-Andersen to Copenhagenize Master Class, 23, 24, 25 June, Copenhagen, Copenhagenize Design Company.

Coman, J. 2015. "How the Nordic Far-Right Has Stolen the Left's Ground on Welfare." *The Observer*, 26 July.

Conservative Party of Copenhagen. 2017a. *Work Plan and Vision for Copenhagen.* Copenhagen, Conservative Party of Copenhagen. Available at https://kbh.konservative.dk/p olitik/valgprogram-2017/ (accessed 28 July 2018).

Conservative Party of Copenhagen. 2017b. *Transport Policies.* Copenhagen, Conservative Party of Copenhagen. Available at https://konservative.dk/politik/politikomraader/transp ortpolitik/?cn-reloaded=1 (accessed 28 July 2018).

Cramer, K. 2016. *The Politics of Resentment: Rural Consciousness in Wisconsin and the Rise of Scott Walker.* Chicago, University of Chicago Press.

Culver, G. 2018. "Bike Helmets: A Dangerous Fixation? On the Bike Helmet's Place in the Cycling Safety Discourse in the United States." *Applied Mobilities*, DOI: doi:10.1080/ 23800127.2018.1432088.

Danish Cycling Federation. 2011. *Highlights from the History of the Cyclists Federation.* Copenhagen, Danish Cycling Federation. Available at www.cyklistforbundet.dk/Om-os/ Om-organisationen/Historien/nedslag (accessed 4 July 2018).

Danish Ministry of Economy and Finance. 2018. *How Communes Raise Taxes.* Copenhagen, Ministry of Economy and Finance. Available at https://oim.dk/nyheder/ nyhedsarkiv/2018/okt/kommuner-haever-skatten-det-udloeser-straf/ (accessed 28 July 2018).

Danish Ministry of Transport. 2018. *The Car Will Have a Greater Role in the Future.* Copenhagen, Ministry of Transport. Available at www.trm.dk/da/nyheder/2018/ole-birk-ole sen-bilen-faar-en-stoerre-rolle-i-fremtiden (accessed 27 July 2018).

Danish Parliament. 2016. *Future of Cycling Policy Summit (Fremtidens Cykelpolitik).* 29 August. Copenhagen, Danish Parliament.

Danish Road Directorate. 2017. *Driving Forces: Why Is Road Traffic Growing in Denmark?* Copenhagen: Danish Road Directorate.

Socialdemokratiet. 2017. *Security and Freedom: 2017 Work Program.* Copenhagen, Socialdemokratiet.

Emanuel, M. 2016. "Copenhagen, Denmark: Branding the Cycling City." In R. Oldenziel, M. Emanuel, A. A. de la Bruheze, and F. Veraart, eds., *Cycling Cities: The European Experience*. Eindhoven, Foundation for the History of Technology.

Emmenegger, P., Kvist, J., Marx, P., and Petersen, K. 2015. "Three Worlds of Welfare Capitalism: The Making of a Classic." *Journal of European Social Policy*, 25: 3–10.

Enhedslisten. 2018a. *Transport*. Copenhagen, Enhedslisten. Available at https://kbh.enhed slisten.dk/politik/transport-og-tilgaengelighed/ (accessed 29 July 2018).

Enhedslisten. 2018b. *The Red-Green Alliance* (English version). Copenhagen, Enhedslisten. Available at http://org.enhedslisten.dk/content/red-green-alliance (accessed 4 January 2019).

Enhedslisten. 2018c. *Political Program for Copenhagen 2018–2021*. Copenhagen, Enhedslisten. Available at https://kbh.enhedslisten.dk/wp-content/uploads/sites/30/2017/06/Enhedslis ten_Kobenhavn_Politisk-P-rogram-2018-21_A5_v2-korrektur.pdf (accessed 29 July 2018).

Esping-Andersen, G. 1990. "The Three Political Economies of the Welfare State." *International Journal of Sociology*, 20: 92–123.

Fitzmaurice, J. 1981. *Politics in Denmark*. New York, St. Martin's Press.

Forchtner, B. and Kølvraa, C. 2015. "The Nature of Nationalism: Populist Radical Right Parties on Countryside and Climate." *Nature & Culture*, 10: 199–224.

Freudendal-Pedersen, M. 2015. "Cyclists as Part of the City's Organism: Structural Stories on Cycling in Copenhagen." *City and Society*, 27: 30–50.

Gössling, S. 2013. "Urban Transport Transitions: Copenhagen, City of Cyclists." *Journal of Transport Geography*, 33: 196–206.

Hagemeister, M. L. and Jensen, C. N. 2018. "Great Day for the Capital: Politicians Praise New Gigaproject in Copenhagen." Danish Broadcasting Corporation, 5 October.

Hansen, A. L., Thor Andersen, H., and Clark, E. 2001. "Creative Copenhagen: Globalization, Urban Governance and Social Change." *European Planning Studies*, 9: 851–869.

Harvey, D. 1996. *Justice, Nature, and the Geography of Difference*. Malden, MA: Blackwell.

Harvey, D. 2008. "The Right to the City." *New Left Review*, 53: 23–41.

Harvey, D. 2012. *Rebel Cities: From the Right to the City to the Urban Revolution*. New York, Verso.

Henderson, J. 2006. "Secessionist Automobility: Racism, Anti-Urbanism, and the Spatial Politics of Automobility in Atlanta, Georgia." *International Journal of Urban and Regional Research*, 30: 293–307.

Henderson, J. 2013. *Street Fight: The Politics of Mobility in San Francisco*. Amherst, University of Massachusetts Press.

Henderson, J. 2018. "Google Buses and Uber Cars: The Politics of Tech Mobility and the Future Urban Liveability." In K. Ward, A. Jonas, D. Wilson, and B. Miller, eds., *The Routledge Handbook on Spaces of Urban Politics*. Abingdon and New York, Routledge.

Jyllands-Posten. 2016. "Danish People's Party." *Jyllands-Posten*, 7 September.

Keman, H. 2017. *Social Democracy: A Comparative Account of the Left-Wing Party Family*. London, Routledge.

Kimmelman, M. 2018. "France's Yellow Vests Reveal a Crisis of Mobility in All Its Forms." *New York Times*, 20 December.

Kosiara-Pedersen, K. 2012. "The 2011 Danish Parliamentary Election: A Very New Government." *West European Politics*, 35: 415–424.

Kuttner, R. 2008. "The Copenhagen Consensus." *Foreign Affairs*, 87(2): 79-94.

Larsen, H. G. and Lund Hansen, A. 2008. "Gentrification: Gentle or Traumatic? Urban Renewal Policies and Socioeconomic Transformation in Copenhagen." *Urban Studies*, 45: 2429–2448.

Larsen, H. G. and Lund Hansen, A. 2015. "Commidifying the Danish Housing Commons." *Geografiska Annaler: Series B, Human Geography*, 97: 263–274.

Leonard, M. 2015. "Why Even Scandinavia Is Moving to the Right." *New Statesman*, 144: 13–14.

Liberal Alliance. 2018a. *Remove the Registration Tax*. Copenhagen, Liberal Alliance. Available at www.liberalalliance.dk/politik/fjern-registreringsafgiften/ (accessed 29 July 2018).

Liberal Alliance. 2018b. *Principal Program and Work Program*. Copenhagen, Liberal Alliance. Available at www.liberalalliance.dk/principprogram-og-arbejdsprogram/ (accessed 29 July 2018).

Liberal Alliance. 2018c. "Old Charges on New Cars." *Liberal Alliance News Magazine Spring 2018*. Copenhagen, Liberal Alliance.

Lidegaard, B. 2009. *A Short History of Denmark in the Twentieth Century*. Copenhagen, Gyldendal.

Radikale Venstre. 2017. *Supplementary Environmental Policy Program for Radical Left Copenhagen*. Copenhagen, Radikale Venstre.

Radikale Venstre. 2018. *Radikale Venstres historie*. Copenhagen, Radikale Venstre. Available at www.radikale.dk/content/radikale-venstres-historie (accessed 30 July 2018).

Rasmussen, J. B. 2014. "Voter Change: Here Comes the Half Million New DF Voters." *Berlingske*, 25 May.

Sedgwick, M. 2013. "Something Varied in the State of Denmark: Neo-nationalism, Anti-Islamic Activism, and Street-level Thuggery." *Politics, Religion & Ideology*, 14: 208–233.

Shoup, D. 2005. *The High Cost of Free Parking*. Chicago, Planners Press, The American Planning Association.

Skrydstrupog, I., and Frandsen, L. 2012. "Wild Battle for the Payment Ring." *B. T.*, 8 January.

Smith, N. 1996. *The New Urban Frontier: Gentrification and the Revanchist City*. London, Routledge.

Social Democrats in Copenhagen. 2017. *2017 Voter Program in Copenhagen*. Copenhagen, Social Democratic Party. Available at www.socialdemokratiet.dk/media/6107/valgprogram-2017-web.pdf (accessed 15 December 2018).

Socialistisk Folkeparti. 2018. *Better and Cheaper Public Transport*. Copenhagen, Socialistisk Folkeparti. Available at https://sf.dk/det-vil-vi/transport/ (accessed 29 July 2018).

Sosa López, O. and Montero, S. 2018. "Expert-Citizens: Producing and Contesting Sustainable Mobility Policy in Mexican Cities." *Journal of Transport Geography*, 67: 137–144.

Verlinghieri, E. and Venturini, F. 2018. "Exploring the Right to Mobility through the 2013 Mobilizations in Rio de Janeiro." *Journal of Transport Geography*, 67: 126–136.

Watts. 2018. "Macron's U-turn on Eco-Tax Rise Gives Green Lobby Fuel for Thought." *The Guardian*, 4 December.

Winther, B. 2017. "Conservatives Have Governed Frederiksberg for 108 Years: But Now It Can Soon Be Over." *Berlingsk*, 31 October.

Venstre Copenhagen. 2017. *Fremtidens København (Future of Copenhagen)*. Copenhagen, Venstre Copenhagen.

Venstre. 2018. *The Right to Live Your Own Life*. Copenhagen, Venstre. Available at www.venstre.dk/politik/principprogram/retten-til-eget-liv (accessed 29 July 2018).

Interviews held by Jason Henderson from 2015 to 2017 with transportation experts in Copenhagen

City Transport Executive (1), 2016.
City Transport Planner (1), 2016.
City Transport Planner (5), 2016.

Transport Advocate (1), 2016.
Transport Consultant, 2015.
Transport Consultant (2), 2016.
Transport Consultant (3), 2016.
Transport Scholar (1), 2016.

5

HOW MANY CARS IN THE CITY?

The Copenhagen toll ring debate

Introduction: how many cars?

In early 2012 Copenhagen, after over a decade of study and debate, was unable to implement a "Bompengering" (toll ring), a type of car restraint meant to discourage car use by charging a fee to drive into the center of the city. The toll ring was meant to limit the number of private cars entering Copenhagen from the suburbs and to dissuade city residents from bringing new cars into the city.

In Copenhagen the toll ring had political traction because suburban cars were (and continue to be) the main source of car traffic in the city (City Transport Planner (6), 2016). On an average weekday 280,000 cars and trucks drove on Copenhagen's streets (City of Copenhagen, 2017; City Transport Planner (6), 2016). Of these, 170,000 cars and trucks entered the city from beyond the municipal borders, while 110,000 originated from within the city or crossed through the city.

On crosstown arteries like the H. C. Andersens Boulevard or the northern and southern segments of the Ring 2, car volumes were high and congestion hot-spots around the city amplified stress for cyclists on the adjacent cycle tracks. Resuscitating arguments in favor of a toll ring in 2018, Copenhagen's Alternative Party pointed out that while 9 percent of Copenhagen residents used cars within the city, 66 percent of Copenhagen street space was dedicated to cars (City of Copenhagen, 2018). This figure, of course, did not include all the suburban cars entering the city, but the point was that Copenhageners were overrun with other people's cars.

The debate over the toll ring in Copenhagen was a proxy for the question: How many cars should be in the city? Fundamentally, for Copenhagen's Left/Progressive politics of mobility, there should be fewer cars and more car-free spaces. As a car restraint strategy, the toll ring would help to tackle capacity issues in Copenhagen's cycling network because congestion pricing would discourage car use, thus opening up more street space that could then be reallocated for the purpose of widening cycle tracks. This would relieve stress for current cyclists and make more welcoming spaces for more cyclists. The toll ring would also simultaneously help to open up street space for separated bus lanes on key arterials without interfering with cycling. This is important because, as described in Chapter 3, in Copenhagen bus passengers and cyclists come into conflict at bus stops located along cycle tracks.

Between 2006 and 2012 the toll ring was vetted and debated as Copenhagen's city planners crafted the bicycle strategy and the goal of a 50 percent bicycle share for commuting and education trips. The Copenhagen 2025 Climate Plan, globally recognized and celebrated as one of the most far-reaching city-scale efforts to mitigate greenhouse gas emissions, was adopted in 2012 and included a goal to reduce driving from 33 percent to 25 percent of all trips within and around Copenhagen. The toll ring proposal was the springboard for both the bicycle strategy and the climate plan, and especially for creating spaces so that 80 percent of the cycle track network could be upgraded to the three-lane-wide Plus-Net system.

The politics of Copenhagen's toll ring was an extension of the politics of the car in Denmark. Copenhagen's Left/Progressive political establishment supported the toll ring because it promised car restraint. In a departure from their historic ambivalence towards the car, Copenhagen's Social Democratic leadership initially supported the toll ring. The Right/Conservatives opposed the toll ring, instead essentializing the car and objecting to car restraint. Somewhat perplexing, because pricing is a market-oriented approach to mobility, the Neoliberals, including business groups and the Liberal party, were critical of the Copenhagen planners' proposal and were either silent or vocally opposed to it.

At first, the national government, led by the Liberals during the formulation of the toll ring proposal (2006–2011), refused to enable Copenhagen. The Social Democrats ousted the Liberals in 2011 and assumed governance of Denmark with a coalition of parties to the left of the Social Democrats, including the Socialist People's Party and the Red-Green Alliance. The Social Democratic Party campaigned in the 2011 election with promises to approve the toll ring for Copenhagen. Yet once in power the Social Democrats fumbled and ultimately scuttled Copenhagen's toll ring.

Understanding Copenhagen's experience with the toll ring proposal is worthwhile for situating and contextualizing debates about car restraint around the world. Surprisingly little is known beyond Denmark's borders about this contentious debate in Copenhagen, which was at times akin to an emotionally charged and vitriolic minefield. Again, this makes Copenhagen's politics of mobility more—not less—like the politics of mobility in peer cities grappling with the same issues. At around the same time that Copenhagen vetted the toll ring, so too did New York and San Francisco in the United States. Like Copenhagen, both proposals floundered and died, and both cities have fallen flat on grappling with the car. Experiences among cities in other parts of Europe and Latin America paralleled Copenhagen. In London and Stockholm, where congestion pricing was successfully adopted, similar political differences between Left/Progressives, Neo-liberals, and Right/Conservatives could be observed, albeit with different outcomes.

Controversy surrounding Copenhagen's toll ring proposal leads to ideas which might appeal to Copenhageners including, with hindsight, the repackaging and implementation of congestion pricing as a car restraint policy. This is not to suggest that Copenhagen's toll ring proposal was itself flawed or inadequate as a car restraint policy—it was mostly well thought out—but rather that understanding the political dimension matters. Copenhagen's experience may also signal that some other car restraint policies, such as car-free centers or more intensive congestion management using new technology, could be more politically viable and practical.

Congestion pricing around the world

Before examining Copenhagen's toll ring debate, we describe the concept of congestion pricing in more detail in order to contextualize toll rings within the broader urban transportation milieu. Toll rings, or cordon-based congestion pricing, involve charging car drivers a fee to enter certain areas of cities during times of peak demand such as weekday commuting hours. Experience in London and Stockholm reveals the merits of toll rings.

After implementing a toll ring in 2003, bus delay in central London declined substantially and ridership increased, while congestion decreased immediately by 25 percent, later levelling off to a 10 percent decrease in congestion (United States Department of Transportation, 2011). Cycling levels in London also increased, and more space was allocated for cycling. Many pre-toll car drivers reportedly switched to public transit or cycling, and there was no significant displacement of traffic to nearby roads outside the toll zone. As of 2018 congestion was thought to be back to pre-toll levels, but this accompanied a reallocation of streets for cycling and buses, and was also accompanied by population and job growth in London. In summary, there is now more space in London for cycling and buses, and if the £11.50 (approximately US $14.00) toll had not been implemented, congestion would be much worse.

Impacts in Stockholm (implemented in 2006–2007) were positive as well. Upwards of 100,000 cars were removed from Stockholm's urban core and public transit ridership increased (City of Stockholm Traffic Administration, 2009). Bypass highways experienced modest increases in traffic, but this was partly due to expected increases in background traffic accompanying regional growth (Börjesson, 2012). With the toll ring in place Stockholm experienced a bicycle renaissance and slowly re-established its cycling infrastructure, and the mode share of cycling rebounded to almost 8 percent (up from 1–2 percent before the toll). At first glance, and at least initially, London and Stockholm successfully implemented and maintained toll rings.

The first toll ring to be implemented was in Singapore in 1975, and it was expanded in the late 1990s following the invention and installation of digital cameras and scanning technology. Automatic license plate recognition, online payment schemes, and more recently smartphone apps invigorated interest in tolling around the world, and in the 2000s Singapore, joined by London and Stockholm, was followed by some less robust variations in Gothenberg, Milan, Oslo, and Rome. With the advent of "connected cars" (a precursor to driverless cars) there is more attention for using global positioning systems and sensors to deploy tolling.

A toll ring, also known as cordon-based congestion pricing, is one of five different kinds of congestion and roadway pricing schemes currently deployed around the world. Four other pricing schemes include distance-based, area-wide roadway charges, such as the German toll imposed on trucks using the Autobahns (motorways), and which was recently tested in Oregon, USA. In this scheme cars or trucks are charged a fee for each mile or kilometer traveled. Variably priced express lanes are special toll lanes usually constructed in the medians of existing highways. These can be found in Southern California, the San Francisco Bay Area, and in greater Washington, DC. Fully tolled roadway segments or bridges (including the Bay Bridge in San Francisco and toll roads in the Texan cities of Austin, Dallas, and Houston) have a longer history but have recently shifted to variable pricing during peak periods. Finally, in some cases car parking charges are a form of congestion pricing, with the most widely recognized being San Francisco's variably priced curbside parking.

From the beginning toll rings were about car restraint. One of the earliest champions of congestion pricing, Columbia University professor William Vickery, framed roads as a scarce resource and pointedly described cars in New York as a "space-hungry, monstrous, rubber-shod sacred cows" (Vickery, 1963). Congestion and roadway pricing has been especially encouraged by many urban planning academics and professionals in the United States who believe that it has great potential for greening mobility and making cities more livable (King et al., 2007). Yet putting academic theory to practice has been difficult in many cases, and the politics is tough to overcome.

Cordon-based congestion pricing—the toll ring concept—was proposed and rejected in Copenhagen, as well as after intense debate in New York in 2008, and with lesser but still contentious politics in San Francisco in 2010. As of early 2019 no such scheme had been successfully implemented in North America, although New York, Seattle, and Vancouver were considering proposals, with New York's closest to realization in 2019. In the United Kingdom, London is the exception. Edinburgh and Manchester debated and rejected the introduction of toll rings. No German cities have cordon-based congestion pricing although such ideas have been floated in Cologne, for example. In France, the national government briefly considered passing legislation to allow urban tolling, but withdrew the proposal following the outbreak of the "yellow vest" protests in late 2018. In South America, the governments in Bogotá and São Paulo briefly pondered and then declined toll rings despite chronic congestion and car-related pollution (Mahendra, 2008). In other cases, such as in Colorado, USA, and in several Chinese cities, road pricing was considered but not adopted. Even in London, the geographic extent of the cordon congestion zone was challenged and reduced in size after the city's Conservative Mayor, Boris Johnson, blocked the expansion in 2011.

Toll rings, as a form of congestion pricing, especially foregrounds a Neoliberal politics of mobility. In the United States, enthusiasm for the concept is decidedly neoliberal in that, according to the United States Federal Highway Administration

(2011), it "harnesses the power of the market," and resembles tactics of private companies, such as peak charging by airlines, hotels, rental cars, and most recently, "surge pricing" by transportation network companies like Uber. Thus, roadway pricing is a commodification of road space, fitting the neoliberal logic among free market advocates that roads are a "commons" being inefficiently allocated.

Road pricing resonates in Right/Conservative politics of mobility as well, but only if the toll revenues are directed to benefit car drivers and in conjunction with privatization schemes. In Southern California and Texas, regions with strong pro-car politics of mobility, right-wing think tanks champion tolling for new highway expansion. For the Right/Conservatives, the rational and efficient management of the commons is not the main goal, but rather expansion of the girth of the commons, in the form of more and more roads. This might work in sprawling Texas or Southern California where there is less concern about the spatial consumption of the car and roads, but not in most world cities.

Chronopoulos (2012), reviewing toll ring schemes in London and Stockholm, and cordon pricing proposed in New York in 2008, notes that congestion pricing was controversial for many Left/Progressives because of the neoliberal dimension. In each of these cities, car traffic burdened the rich and middle class equally in that all car drivers, regardless of income, experience delays. (The lower class in these cities was presumably less affected because they did not drive to begin with.) Wealthy drivers (or passengers in hired cars) were delayed by middle-class drivers, and vice versa. With the introduction of congestion pricing in spatially constrained cities the wealthy could buy mobility privileges and shake out or bypass some of the middle-income drivers.

In this vein congestion pricing is a spatial mobility policy that privileges affluent drivers and prices out the less affluent, but they are in fact mostly middle-class drivers. To make congestion pricing politically possible in cities with large left-leaning and middle-class constituents, redistributive approaches directing revenue to transit are considered necessary (King et al., 2007; Chronopoulos, 2012). If funds generated by tolls are redistributed and used to underwrite public transit and cycling infrastructure, many goals important to the Left/Progressive politics of mobility might be achieved. Cities such as London and Stockholm, for example, steer funds raised by congestion pricing towards public transit, cycling, and improvements for pedestrians.

Still, to some on the left the commodification of streets, even with redistribution, remains exclusive, elitist, and unfair (Ross, 2006; Chronopoulos, 2012). Left/progressive scholars such as David Harvey (2012, pp. 74–75) acknowledge that car-congested streets are a major problem in cities, and that when streets are clogged with private cars they are not a commons for everyone anymore. Yet to Harvey congestion pricing simply raises the price of entry to the more exclusive commons and does not bring about a return of streets as a commons for all (here he is critiquing ecologist Garret Hardin's 1968 "Tragedy of the Commons" thesis).

From the left, the tragedy of the commons is not that the street is publicly owned and overrun with cars, but that the cars that ruin the commons are privately owned. The problem is not the public street, but the private car on the street. Cycling and public transit might be a type of reintroduction and recovery of the street as commons, but only if everyone—not just the wealthy—can afford to enter the congestion zone, or pay the rent for housing within or nearby the congestion zone. This implies that private cars must be almost completely removed from the commons instead.

Copenhagen's toll ring proposal partially addressed this Left/Progressive "right to the city" approach to the commons because the aim was not to simply make it easier for wealthier motorists. Instead, redistribution included (albeit that it was vague and generalized in the plan) an intention to reallocate street space away from cars and towards cycling and public transit. While not eliminating all private cars, Copenhagen's toll ring was intended as a means of car restraint and of returning part of the commons to people, not cars. It was an extension of the idea of the right to a green mobility city, and Copenhagen's Left/Progressives (the Red-Green Alliance, and the Socialist People's Party) along with Social Democrats and the Social Liberals, supported the toll ring concept with this understanding (the Alternative had not yet established itself as a party when congestion pricing was debated).

A brief history of Copenhagen's toll ring debate

In Copenhagen, the idea of a toll ring with leftist undertones goes back to the 1970s. Responding to the oil crisis and related environmental movement, attempts to introduce tolling were made by smaller left-wing parties in the municipality of Copenhagen but were rejected without study (Homann-Jespersen *et al.*, 2001). In the 1990s the Left/Progressive Socialist People's Party and the Left-Socialist Party (the precursor to the Red-Green Alliance) promoted the tighter regulation of cars in Copenhagen and considered tolling as a redistributive policy. In 1993 Copenhagen's Traffic and Environment Plan for Copenhagen, with a planning horizon of 2005, called for "future regulatory instruments" to limit cars and this included studying a toll ring and payment stations around the city (ibid.).

For Copenhagen's Left/Progressives this made sense. Few Copenhageners owned cars and the city was in the throes of deindustrialization and decline. Most residents were lower-income students, older pensioners, and the unemployed and the municipality of Copenhagen had one of the lowest rates of car ownership of all developed world cities (185 cars per 1,000 residents in 1994) (Cervero, 1998; City Transport Planners (2 and 3), 2016). Tolling wealthier suburban car owners (and home owners) meant that the socialists were redistributing from the bourgeoisie while recapturing some of the commons and reallocating city street space for cycling and public transit rather than for the car.

The geography of car ownership mattered. By the end of the 1990s 42 percent of households within 6 kilometers of Copenhagen's center had access to a car, while in the inner ring suburbs 70 percent of households had a car (Naess, 2006, citing a travel survey data from 2000). This climbed to 82 percent in the outer suburbs of Copenhagen (ibid.). Ironically, despite origins and theoretical underpinnings as a market-based, neoliberal concept, Copenhagen's Left/Progressives sought to make suburban private car drivers pay to enter the city with a modified market-based approach.

Copenhagen's Left/Progressives were not alone in considering tolling. The neighboring Scandinavian capital Stockholm had considered tolling as far back as the 1970s and was studying it in 1992 (Oldenziel *et al.*, 2016). Left-wing parties stood behind the schemes. London was studying congestion pricing through the 1990s and it was implemented in 2003 by a socialist mayor. In New York and San Francisco, many (but not all) Left/Progressives in each city backed the concept.

Similarly to those in Copenhagen, London and Stockholm Left/Progressives had to convince their national governments (who were more Right/Conservative) to enable the municipality to implement toll rings (Singapore, a city-state, did not have this political barrier). Ostensibly the legal argument in each case was that the national government must approve new taxes, or at least give a municipality or region the right to tax. London's enabling legislation to levy a tax came with the creation of the Greater London Council, and Stockholm's ability to implement tolls was part of a 2006 agreement that included revenue sharing with the national government.

In New York and San Francisco, the cities' respective state governments had statutory authority over taxation, and before each city could implement congestion pricing, state enabling legislation was required. Copenhagen, New York, and San Francisco were all unable to implement a toll ring, partly owing to Right/Conservative defense of automobility beyond the municipal borders, and in the two US cities, lack of clarity regarding the redistributive aspect of the schemes (in 2019 New York's proposal was more clearly tied to improving the subway system).

To be sure, the politics of mobility for the Danish Right/Conservatives was open to tolling in some form and various studies with a more neoliberal bent were commissioned. Nationwide road pricing was considered (and is still promoted), with the idea that it would make it easier and faster to drive for those willing to pay, and that revenues would be aimed at roadway improvements. For the Right/Conservatives, the national car tax would be reduced to offset tolls. This scheme of offsetting road pricing with reduced car taxes became the congestion pricing framework adopted by the Right/Conservatives and the Neoliberals (in 2018 the Liberals and the Liberal Alliance sought the reduction of car taxes with the idea that car ownership should be taxed less, but perhaps car use taxed more).

By the late 1990s and early 2000s there was widespread political support for tolling in Denmark, but with strong ideological differences over the purpose of tolling. On the left, tolling was about car restraint, redistribution, and the environment. On the right, tolling was about congestion management and an opportunity to reduce the car tax and stimulate more car ownership. There were left–right fissures over how revenue from tolls should be spent. The left held redistributive positions and would have directed revenue to public transit or cycling. The right called for more roads consistent with their right-wing brethren in the United States, where political support for pricing hinged on it being a true user fee to benefit motorists—not a redistributive tax. For the Right/Conservatives and also the Danish Social Democrats, a measure such as a toll ring would only be acceptable if pricing actually made it easier to drive, at least for those willing to pay the toll.

Congestion pricing as redistributive car restraint was debated by Copenhagen's City Council in the late 1990s but was only supported by two left-wing parties (Transport Scholar (1), 2016). For the Socialist People's Party and the Red-Green Alliance, the purpose of congestion pricing was to limit the car (Homann-Jespersen et al., 2001). The Social Liberals, the small, neoliberal-progressive party, also endorsed congestion pricing. The Social Democrats, which outnumbered these parties in terms of membership, objected to the toll ring.

Given the Social Democrats' historical ambivalence towards the politics of the car, the party's adversity towards the toll ring was predictable and resonated with tactics to retain suburban middle-class car owners. Since either the Social Democrats or the Liberals were in government at the national level during the congestion pricing debates, Copenhagen's Left/Progressives could not get the enabling legislation adopted. The national governments of the 1990s, both the Social Democrats and the Liberals, maintained a vice-like grip on the city's aspirations for car restraint and redistributive policies.

In sum, during the 1990s it was politically desirable to regulate and price the car in the leftist urban core in Copenhagen, whereas the higher car ownership in the suburbs made it more difficult to secure political support. Outside Copenhagen most households owned a car, and this meant that in suburban electoral politics, pandering to car owners was routine. A decade later a similar politics of car ownership helped to explain why car-lite Manhattan had high rates of political support for the 2008 congestion pricing proposal in New York. In outer parts of the boroughs like Queens, and in Long Island and upstate suburbs, higher car ownership rates undergirded political opposition to New York's proposal. Similar political geographies characterized the San Francisco congestion pricing proposal, which was beaten back by suburban politicians in 2010.

In 2001 Copenhagen's Social Democrats tilted back towards the Left/Progressive position, and entertained the idea of using a toll ring to regulate cars and raise revenue for the city (Transport Scholar (1), 2016). It seemed for a moment that Copenhagen might possibly get the national enabling legislation it needed. At this point and for the first time the majority in the Copenhagen city council supported congestion pricing in the form of a cordon or toll ring, with the location of the ring to be determined through more study. A new 2001 study on regional traffic and the future of the region listed congestion pricing as a promising solution.

Led by the Social Democrats both the city authorities and national government agreed on further detailed study of cordon congestion pricing. To remind readers, this was the same time that London also vetted congestion pricing and the city's left-wing Mayor Ken Livingstone started to champion the concept for London. The British government accordingly granted London the right to implement congestion pricing.

Yet in 2001 Danish national politics shifted to the right again, with a new Liberal party-led government that was hostile towards the municipality and intent on thwarting Copenhagen's political influence on the rest of the country, including the proposal to introduce congestion pricing. For almost a decade the congestion pricing scheme for Copenhagen languished, although not without further study and political maneuvering.

The Right/Conservative bloc refused to consider congestion pricing and characterized the concept as an unfair tax burden on Danish car drivers outside Copenhagen (Transport Scholar (1), 2016). Ironically the Liberals embraced an array of neoliberal privatization and deregulation policies during its ten-year tenure in the early 2000s but not congestion pricing. The Liberal party and its right-wing coalition partners weakened the Ministry of the Environment, which had taken a very strong green mobility stance on limiting the car during the 1990s, and relaxed an important land use planning law requiring that new development must be proximate to railway stations (we return to this topic in Chapter 7). Making the law advisory-only, the Liberals hobbled what remained of Copenhagen's revered Finger Plan, and encouraged car-oriented sprawl development around Copenhagen.

The Liberal party was also invested in Copenhagen's economic recovery which was underway by the early 2000s. It was loath to introduce tolls just as the city began to see new redevelopment schemes around the harbor come online, which were attracting higher-income residents and more jobs (City Finance Executive, 2016). Car restraint in the form of pricing was seen as working at cross-purposes to redevelopment policies. For some Social Democrats as well, the proposed toll would also undercut this growth, and thus impact revenue collection to underwrite social welfare in the city.

Finally, the Liberals also reorganized the administrative regions of Denmark, bifurcating Copenhagen and the suburbs from regional cooperation on planning and transportation even further. These deregulatory and privatization schemes were consistent with Neoliberal and Right/Conservative policy, yet ironically when it came to pricing and the car, pricing was not acceptable. This contradiction has been noted by sustainable mobility scholars like Newman and Kenworthy (2015) who remark that when it comes to cars and roads, the market is subsumed to socialist management, despite most other aspects of advanced capitalist nations deploying market approaches in order to manage resources.

With no support from the national government, Copenhagen still pursued congestion pricing. In 2006 the duo comprising the newly elected Social Democratic Mayor, Ritt Bjerregaard, and the Social Liberal Klaus Bondham, deputy mayor of the technical environment department, kept the toll ring idea alive in Copenhagen. As these center-left politicians promoted Copenhagen as an iconic world cycling city, they also directed city staff to study a toll ring. Bjerregaard promised congestion pricing in her 2006 campaign for mayor and recognized the success of Livingstone in London three years earlier. Livingstone survived a re-election campaign in 2004 and vowed to expand London's tolling zone, which was accomplished by 2007 (and was then rescinded by the right-wing government in 2011).

Instead of economic decline, in 2006 Copenhagen was now in the midst of reurbanization, economic boom, and population growth. Car congestion increased. Between 2000 and 2004 congestion increased by 20 percent in the municipality of

Copenhagen and by 35 percent in the greater Copenhagen region (City of Copenhagen, 2009). Traffic congestion was increasing by 10 percent annually, and commuting times in the region were lengthening, thereby threating quality of life and a diseconomy.

Elsewhere the Swedish government granted Stockholm the right to experiment with a congestion pricing pilot scheme, and New York also began to consider one. San Francisco was awarded federal grants to investigate various congestion pricing and parking pricing schemes as well. In hindsight it seemed like the opportune time for Copenhagen to amplify the idea, and City Hall directed planners to draw up plans and to collaborate with the suburbs of Copenhagen in order to develop political consensus.

The results of the political support in Copenhagen was the *Forum of Municipalities Report: Congestion Charging in the Greater Copenhagen Area*, led by Copenhagen's authorities and inclusive of sixteen neighboring suburbs (Forum of Municipalities, 2008; City Transport Planner (5), 2016). City-suburb collaboration included regional revenue sharing and plans to coordinate new bus lines and public transit expansion that would be frontloaded with the toll ring, as well as discussion of traffic patterns along the cordon edge.

The report proposed a 25 kroner peak toll (roughly US \$5.25 in June 2008, or \$4.00 in 2018. For comparison, London's toll was £5.00 in 2003, and by 2018 it was £11.50 per day (approximately \$7.00 and \$14.00, respectively); New York's most recent 2018 proposal was between \$11.00 and \$12.00), reduced off-peak rates, weekends, and holidays, and free overnight (Forum of Municipalities, 2008). The toll would capture any car entering or leaving Copenhagen, so it would not only discourage inbound driving but would capture the rapidly growing reverse commuting pattern of new residents of Copenhagen driving to work in the far-flung suburbs.

Two variants of tolling were considered. The first was to be by way of a ring of electronic cameras using digital photography (resembling London, Singapore, and Stockholm), the second by providing drivers with toll tags with GPS tracking, triggered when crossing the toll ring. The national government would collect the toll or allow Copenhagen to establish a congestion charging authority to collect the toll, presuming that enabling legislation was in place. This congestion authority would operate the toll ring and distribute funds with a board of directors comprised of local elected officials or proxies.

In either format, Copenhagen would share revenues with the suburbs, with an emphasis on redistributive funding for regional transit infrastructure improvements. Examples included financing for an outer Ring 3 light rail line paralleling the Ring 3 motorway, an Atlanta, USA, or London M25-style beltway around Copenhagen's western suburbs. The scheme would also finance increased capacity and frequency on the regional S-trains, build bicycle infrastructure connecting railway stations and including additional parking for cyclists, intercept park and ride for car drivers, expanded bus capacity in the city of Copenhagen, and potential conversion of motorway emergency lanes into express bus lanes. In the longer-term, potential expansion of the Metro was also considered.

Similar to London's scheme, a "traffic start-up package" would quickly expand buses and build new bicycle infrastructure in advance of the toll ring's deployment. This would be financed by borrowing against future toll revenue. The 2008 *Forum of Municipalities* report considered the optimal location of a toll ring within the S-train circuit ("Ringbanen") on the western side of Copenhagen and the Øresundvej to the east in Amager (see Figure 7.1). Since Copenhagen's Ring 2 on the western edge was just beyond the toll ring border, it would function as a bypass of the toll ring, and would need major advanced preparation to handle bypass traffic.

Fundamentally and most importantly for Copenhagen's aspirations to become an iconic bicycle city, car traffic volumes would drop to holiday volumes—23 percent less than a normal working weekday. Some advocates argued that car use would drop by one-third and that the Brokvarterene and Indre By would be radically transformed with a slightly higher toll rate (Transport Advocates (2), 2016). The equivalent of this much car space (a 23 percent to 33 percent reduction of car traffic meant ample new roadway space), would be reallocated to cycling and bus lanes. An example was already underway on Nørrebrogade, and at this point Copenhagen was promoting the Plus-Net scheme to widen cycle tracks to incorporate three lanes of cycling each way.

The city's bicycle reports and subsequent *Action Plan for Green Mobility* (City of Copenhagen, 2013) suggested that the bicycle mode share for all trips within and out of Copenhagen would rise, pointing to the goal for 50 percent of all trips to places of work and education to be by bicycle. Regional car traffic outside Copenhagen would drop by 4 percent, and significantly would stop growing. Motor Ring 3, the Atlanta/London M25-style beltway around western Copenhagen, would however, experience increased traffic. Polling in 2008 showed a 65 percent to 75 percent approval rating in Copenhagen for congestion pricing, and the political calculus seemed to favor the toll ring once and for all.

The backdrop for the toll ring proposal was an exciting time for Copenhagen's green mobility image. In 2009 Copenhagen City Council adopted climate goals which would eventually become the celebrated Copenhagen 2025 Climate Plan which sought to reduce all car trips from 33 percent to 25 percent (City of Copenhagen, 2012). Copenhagen was on the world stage for the Climate Change Conference of the Parties (COP15) in December 2009, which was attended by US President Barack Obama and other world leaders, and the political climate in Denmark was ripe for implementing congestion pricing as a type of climate change mitigation (Transport Scholar (1) 2016). Copenhagen was aggressively branding the bicycle city image as well, and hosted the "Velo Cities" conference attended by cycling advocates from around the world, but especially North America. Revered in urbanist circles, Jan Gehl (2010) chimed in and stated that Copenhagen needed a London-style congestion charge.

During 2009 and 2010 the city fine-tuned the proposed toll ring, with internal and localized debates over the exact alignment of the toll ring and how to handle potential spillover traffic along Ring 2 (City Transport Planner (5), 2016). Representatives of some outer neighborhoods expressed concern about dividing constituents inside and outside the toll ring border. Some argued that the toll ring would be unfair to parents who drove their children to school and had to cross the cordon. Debates over the inevitability of the car arose, as some argued that car drivers would change their route or time of travel rather than forego driving. Others suggested that Copenhagen did not have enough serious congestion to warrant the proposal in the first place (Rich and Nielsen, 2007). None of these arguments undercut the unified push by Copenhagen but did foreshadow the political obstacles to come.

At the national level the Liberals still stood in the way, despite recently embracing some green policies in the run-up to the 2009 climate summit (Transport Scholar (1), 2016). The Liberal's assessment was that Copenhagen would not need the toll ring if national road pricing was implemented instead (interestingly, justifying the need for some form of congestion pricing). The Neoliberal-Right/Conservative coalition government led by the Liberals proposed distance-based tolling on trucks in 2012 and then cars in 2014, potentially stonewalling Copenhagen's toll ring proposal.

In October 2011 the Social Democrats enjoyed a strong showing in the national election, and having campaigned in support of Copenhagen's toll ring, resumed power and governed Denmark. Copenhagen eagerly awaited the enabling legislation by which the city's green mobility vision could leap forward. The city had reason to be optimistic because during the 2011 national election campaign all the Left/Progressive political parties in Denmark promised to implement Copenhagen's toll ring (Transport Scholar (1), 2016). With a green mobility package central to its party platform, congestion pricing was the foundation of the Social Democratic Party's climate and mobility plan for Copenhagen; the Red-Green Alliance, the Social Liberals, and the Socialist Peoples' Party were aligned behind the toll ring (Transport Consultant, 2015; City Transport Planner (5), 2016; Transport Advocate (2), 2016; City Climate Planner (1), 2016). When the Social Democrats assumed parliamentary, the municipality of Copenhagen lobbied for the toll ring to be enabled, and the Social Democrats agreed to shepherd the plan.

Alas, things quickly unraveled from thereon. Rather than guide through Copenhagen's proposal, the Social Democrats appointed the National Road Directorate, part of the Ministry of Transport Building and Housing, to study a plan of its own. This was not the municipality of Copenhagen's plan, and from the outset it omitted the local consensus and agreement created by Copenhagen and the suburbs. Meanwhile, within the ranks of the Social Democratic Party, skepticism about the toll ring, silenced during the campaign, was exposed, and factions of Social Democrats outside Copenhagen balked at the scheme.

A divisive debate unfolded in Parliament, with critics of congestion pricing claiming that the necessary investments in regional transit had to come first, before car drivers were burdened with pricing. Some also balked at Copenhagen's intention of reallocating street space once the toll was implemented. Proponents of the toll ring pointed out that there was a decent transit system already in place with the S-trains, the Metro was under construction, and the national government under the Liberals had expanded the roads generously. The toll ring was needed to make all of this work better, as well as to underwrite transit expansion (Transport Scholar (1), 2016). Proponents remarked that the opposition appeared to "make perfect the enemy of the good," in that they refused to support car restraint until transit was perfect.

Political support eroded and the Danish Social Democrats abandoned official support for the toll ring in parliament. A hastily thrown-together 150-page report from the Road Directorate was presented to parliament and dismissed within ten minutes, according to witnesses, because it was considered not politically viable (City Transport Planner (5), 2016). After over a decade of planning and consensus building, and local Social Democratic Party buy-in, within a few short months the Social Democrats fumbled and Copenhagen's toll ring proposal was scuttled in February 2012. As one local city official put it, "the Social Democrats could run all the way to the finish line with everyone behind them, sure to win, and then trip" (City Climate Planner (1), 2016).

Next the Social Democrats promised to find money for the transit proposals that were to be funded by the toll ring, and established a Congestion Committee to investigate congestion pricing ideas beyond a toll ring around Copenhagen (Transport Scholar (1), 2016). The Danish Commission on Congestion, as it became known, had twenty-eight members including academic experts, members of parliament, mayors, and representatives of non-governmental organizations. The Commission on Congestion, according to some members, was meant to fail. It deliberately comprised many divergent viewpoints that made it impossible to find consensus about a toll ring.

The politically expedient purpose of the committee was to simply placate any remaining toll ring supporters. At the conclusion, two years later, only a vague national road pricing scheme was proposed, aligning the commission with the Liberals and the neoliberal side of the debate over the car, especially because it also recommended reducing the car tax to offset tolls. With political gridlock and a mobility stalemate, that scheme was also rejected (Transport Advocate and Politician, 2015; Transport Advocate (2), 2016). Said one skeptical observer, "A committee like this is where you park ideas" (City Climate Planner (1), 2016).

For its part, Copenhagen's Left/Progressives continued to press for a toll ring, and the Red-Green Alliance especially so. Meanwhile, in a stunning and narrow upset, the Social Democrats were ousted from power in 2015 and replaced again by a Neoliberal-Right/Conservative governing coalition led by the Liberals. The Liberal party reaffirmed that it had no interest in a toll ring in Copenhagen. Furthermore, the Liberal party returned to its policies of building more roads and shifting funding from the railways and public transit, while also endorsing reduction of the car tax (which was achieved in 2015). Without any mitigating measures, Denmark's Neoliberal and Right/Conservative parties were satisfying their urges to make it easier to own and drive a car, without any semblance of offsets to counterbalance the cuts to the car tax.

Why does Copenhagen not have a toll ring?

Copenhagen's toll ring proposal was fairly well developed and based on sound reasoning, yet it was politically charged and ultimately maligned and defeated. What can be learned from this? Sørensen *et al.*, (2014) argue that successful congestion pricing schemes need a project champion, a clear mandate, and the general public must get something in return. In London Ken Livingstone and in Stockholm Green Party activists provided clear mandates about reducing congestion and emphasizing public transit, and the public got better transit and a nascent cycling renaissance in exchange for toll rings. In Switzerland, fees imposed on trucks using Alpine motorways went towards financing a new railway tunnel and in turn the Alps got less truck traffic, thus benefiting Swiss Alpine enthusiasts as well as the tourism industry (ibid.).

These kinds of tangible infrastructure promises (public transit, cycle tracks, less truck traffic) were also part of Copenhagen's plan, and within the municipality of Copenhagen there was leadership (by the Left/Progressives and by a Social Democratic mayor) and clear policy mandates (such as reducing car trips to 25 percent in the climate plan, and increasing cycling for work and education to 50 percent). The city's bicycle plans also called for more substantial street reallocation on Nørrebrogade, Amagerbrogade, and Vesterbrogade, among other key cycling arteries with capacity problems. Furthermore, the Social Democrats campaigned on the promise of a toll ring, and were voted into power in 2011.

Why, then, was the outcome failure? Here we provide three further insights into the politics of the toll ring, reflecting the broader politics of the car in Copenhagen and Denmark, and consider how these may have combined into a perfect storm of rejection of the toll ring. Our autopsy of the toll ring debate is in no way completely conclusive, but we believe it contributes to a better understanding of Copenhagen's politics of mobility and that can be useful in making a comparison with or perhaps informing the experience of other cities.

Social Democratic party-political calculus

After the 2011 Danish parliamentary elections, the Social Democrats looked towards the next election and counted votes in the suburbs. The party feared that it would lose in the next election cycle because despite increasing its share of the vote and winning at the national level, votes for the Social Democrats actually declined in the Copenhagen suburbs. Social Democratic leaders argued that this was because of the party's support for the toll ring in Copenhagen. Social Democratic mayors in the suburbs were also jittery about the results, and pressed the national party to withdraw support for the toll ring after the election results came in, fearing that they too could be replaced in the next round of municipal elections. Following the election, Copenhagen's suburbs descended into "rioting" as described by one observer (City Finance Executive, 2016), and a huge debate developed over the toll ring. Because of that, the national government, even though it was made up of the same polical party as the city's rulers, was prepared to obstruct the wishes of the city (Transport Consultant (1), 2016).

The Social Democratic leadership calculated that implementing the toll ring would compromise the party's success in the next election. As one insider put it, "the Social Democrats got scared" (City Climate Planner (1), 2016). Put another way, the Social Democrats in the national government betrayed their allies in Copenhagen (Transport Scholar (2), 2016).

The conclusion that Social Democrats lost suburban voters because of the toll ring proposal was disputed by the Red-Green Alliance. The alliance saw its share of the vote increase in suburban Copenhagen during the 2011 election, and likely drew away Social Democratic voters. The Red-Green Alliance strongly endorsed the toll ring proposal, and therefore if voters switched between these two parties, it was perhaps not due to the toll ring (City Transport Planner (5), 2016; City Climate Planner (1), 2016: Transport Consultant, 2015).

Furthermore, Copenhagen's suburbs are not one monolithic political bloc, although rates of motorization are higher throughout the suburbs than in the city, and the suburbs are more conservative than the city (City Climate Planner (1), 2016, Transport Advocate (2), 2016). Copenhagen's northern suburbs are wealthier and are more politically aligned with the Liberals, than with the Social Democrats or the Left/Progressives. The "Gold Coast" or "whiskey belt" along the Strandvej (Beach Road) from Gentofte to Helsingor is the wealthiest part of the country, and there is more political hostility towards the public sector compared to other parts of the region. Here support for road pricing aligns with Neoliberal aims of making it easier and faster for the wealthy to drive, and subsequently reducing the car tax. Copenhagen's toll ring was not the kind of pricing scheme these Neoliberals preferred. A harbor tunnel with high tolls was deemed more suitable, as we will explore in Chapter 7.

To the west and southwest of Copenhagen are more middle- and working-class suburbs inhabited by many traditional Social Democrat voters. These suburban precincts have seen a marked increase in voters shifting to the right-wing Danish People's Party, not because of mobility politics, but because of anti-immigrant and racialized reactions to increased concentrations of non-Danish immigrant groups in this part of the Copenhagen region. There might have been resentment towards Copenhagen's toll ring proposal as well, but race and immigration politics is what steered voters to the Danish People's Party, not the politics of mobility.

Given that a proportion of previously Social Democratic suburban voters shifted to the Red-Green Alliance, which supported the toll ring, and another proportion shifted to the right due to immigration politics, it is less probable that the toll ring factored into declining Social Democratic voter turnout in the suburbs of Copenhagen. Instead, it appears that broader dissatisfaction with the Social Democrats was more salient.

It should also be pointed out that after scuttling the toll ring, the Social Democrats still lost the next national election in 2015. Promising suburban middle-class motorists to defend their interests did not bring Social Democrats victory. It might be that failure to lead, rather than following misleading polls or electoral analysis, is a lesson for the Social Democrats.

That said, post-2011 election analysis showed there was a clear city-suburb divide within the Social Democratic Party. In Copenhagen, the Social Democratic mayor and members of the City Council supported the toll ring, and had widespread support from the other Left/Progressive parties, especially the Red-Green Alliance. If Copenhagen had been able to pursue the toll ring on its own, without the need for national enabling legislation, Social Democrats within the municipality would probably have implemented the toll ring. They would have likely had support from the other Left/Progressive parties, even if there was some opposition from the small Right/Conservative or Neoliberal representation on the City Council.

More bluntly, many national and suburban Social Democratic Party critics of the toll ring simply wanted to defend the growth in car ownership and usage because of the financial connection to social welfare. They warned that congestion pricing would reduce car and fuel sales which the national government depended upon for revenue to support its social welfare programs (Transport Scholar (1), 2016). Car-commuting suburban workers employed in Copenhagen also argued that the toll ring would reduce their incomes and so was tantamount to an income tax increase (City Finance Executive, 2016). Copenhagen would gain revenue through collecting the toll, but suburban workers and the national treasury might lose revenue if driving declined. Unsustainable mobility was tied to economic prosperity and social welfare. If reduced revenue from cars resulted in cuts to social welfare, the Social Democrats would be blamed.

The Social Democrats, whether believing it necessary to preserve voting margins, or because of sincere concern over suburban car drivers, or because of perception that car ownership was necessary for financing social welfare, displayed a lack of political will to implement the toll ring. The sequence of events described here are consistent with the Social Democratic Party's historical ambivalence towards the car and the inclination to conflate and overemphasize middle-class social welfare with car ownership.

Shifting storylines and toll ring border

Contemporary observers of Copenhagen's toll ring debate suggest that mistakes were made by the city of Copenhagen's promotion and presentation of the toll ring (Transport Scholar (1), 2016). One storyline communicated to the public was about reducing congestion and making life easier for drivers who paid the toll. Another narrative concerned car restraint that reallocated space but did not necessarily make it easier to drive. Another narrative emphasized reducing CO_2 emissions, while yet another narrative promoted the toll ring as a revenue-generating necessity. Practically speaking, the Left/Progressive version of the toll ring was all but the first of these things—the Left/Progressives did not seek to make it easier to drive in Copenhagen. Promoters of the toll ring needed discipline to keep the messaging succinct, consistent, and genuine. The messaging unraveled for several reasons.

One "catastrophe in debate," described by a proponent of the toll ring, was that the boundary of the toll ring was not finalized at the time of the Social Democratic Party's return to national governance in 2011. The city suggested that it preferred the Ringbanen toll ring border. But in fact the location of the border remained controversial and depending on where it was located, meant different things to different people (Transport Scholar (1), 2016). Careful planning and structuring of the toll ring scheme was evident, but not the border. The city of Copenhagen preferred the western Ringbanen, which the city concluded was the most popular option, but others said traffic would flow better if the Ring 2 was the western border, and still others suggested that the municipal boundary was more suitable. At some point even the Ring 3 beltway was suggested. The lack of agreement on the exact alignment of the toll ring left the proposal exposed to criticism because it also confused the exact purpose for the toll ring (Transport Advocate (2), 2016).

For example, the underlying intention for Copenhagen's Left/Progressives was not simply to manage congestion, but to reallocate street space to improve cycling and public transit. That is, the toll ring plan was not actually meant to reduce congestion, but to actually further constrain the car by reallocating space away from cars once the toll was in place. The Left/Progressive scheme charged motorists to drive but did not necessarily promise less congestion. It did, however, promise that cyclists would glide past cars on extra-wide cycle tracks and that buses would traverse the city unimpeded.

With car restraint as the goal, this meant that a tighter border was probably more suitable, at the Ringbanen or Ring 2 on Copenhagen's west side. If the goal of the toll ring was greater ease of driving for those willing to pay, street space within the toll ring would still need to be preserved. The wider the geographic area (such as out to the municipal border or further out to Ring 3) the less feasible it became to implement the kind of car restraint envisioned. If a car drove within the border of the toll ring and did not cross the cordon at any point, it was left un-tolled, meaning a wider geographic area could become meaningless (unless distance-based or area-wide tolling was deployed in conjunction with the toll ring).

The Social Democrats at the national level did not buy into the Left/Progressive scheme of car restraint and all of the other main parties, from the neoliberal Liberal party, and the Right/Conservative parties such as the Danish People's Party, protested against reducing car access to Copenhagen, and especially against the toll ring concept (City Climate Planner (1), 2016).

The ambiguity over the extent of reallocating street space and the location of the toll ring border contributed to media embellishment and hyperbole. The media looked into the question as to why people who needed to drive to Copenhagen would be charged, but those who did not need to drive into the city itself but instead drove suburb-to-suburb would be left un-tolled (suggesting that the Ring 3 solution was a fairer option, albeit that a much lower daily fee would be charged). The media sensationalized stories of parents having to drive their children to schools from one side of the toll ring to another, with the result that they were charged for chauffeuring their children. One story involved a single mother with two children being interviewed in an emotionally charged plea for the abolition of the toll ring.

As the media hyped up the opposition and thrived on fear, congestion pricing lost popular support and became bad news. Much of the media attention was based on anecdotal reporting that was emotional and hearsay rather than accurate reporting of the facts (Transport Scholar (1), 2016; City Transport Planner (5), 2016; Transport Consultant, 2015). The media also elevated criticisms that the toll ring was old-fashioned and that road pricing on a regional or national scale would be a better approach, despite that also not coming to fruition. Other expert opinions suggested Copenhagen pursue parking pricing instead of the toll ring, although as we will see in Chapter 6, many of the same ideological barriers stand in the way of parking pricing and management.

Branding of the toll ring was also criticized and inflamed in the media (Transport Scholar (1), 2016). Labeling the congestion charge as a toll or payment ring may have been a mistake because in the media it was compared to a new medieval city wall, a negative connotation suggesting that Copenhagen only sought to keep suburbanites out of the city. Confusing the matter, the city tried to rename the toll ring to a "green" or "sustainable" ring to deflect criticism but by then it was too late. The concept had gained a maligned reputation in the media (Transport Scholar (1), 2016; City Finance Executive, 2016). While perhaps a little disingenuous because the toll ring was both literally an accurate definition and a strong technical solution for achieving Copenhagen's goals of car restraint, perhaps softer green branding and marketing might have disempowered some of the negativity.

The geographical fluctuation over the toll ring border also played into questions about the sense of urgency about the need for congestion pricing. While Copenhagen's Left/Progressives and Social Democrats were arguing that the city needed the toll ring, city planners were boasting that Copenhagen had always managed traffic well, with a long history of traffic management that combined signaling and road space to limit the number of cars in the city to volumes recorded in Copenhagen prior to the contemporary reurbanization boom (City Transport Planner (5), 2016). Since the 1970s city traffic engineers had used signaling to manage new traffic and some in the media and public asked, why was pricing necessary? On average in Copenhagen, cars spent at most 15 minutes in congested conditions. This was considered normal in Denmark and throughout Europe and congestion was sometimes worse outside the city (especially on roads like the Ring 3).

Copenhagen ranked 70th among Europe's most congested cities with a population over 250,000 (City Transport Planner (6), 2016, citing Inrix, 2015). In retrospect, some Copenhagen city planners observed that "Compared to Paris and London, Copenhagen's perception is that it doesn't have a congestion problem" (City Transport Planners (2 and 3), 2016). Critics wondered if the problem was actually serious enough to warrant the implement of a toll ring in Copenhagen.

Class antagonism and the toll ring

In postmortems of congestion pricing schemes that failed to materialize, class resentment among middle-class car owners is frequently identified as a major factor (Chronopoulos, 2012). Populist claims of elitism from both the left and right could be heard in New York and San Francisco in the United States, and in Edinburgh and Manchester in the United Kingdom. In France, the populist "yellow vest" protests in 2018 resulted in the government retreating on congestion pricing proposals. Class antagonism is aggravated by the assumption that congestion pricing is genuinely for the rich (Ross, 2006). The wealthy would continue to drive despite the congestion charge, while many middle-class drivers are forced to switch to public transit or forego the trip into the urban core where they would otherwise like to travel.

Referring back to Harvey's (2012) critique of tolling the commons, and specifically congestion pricing in cities, the gentrification of cities like Copenhagen aggravates this kind of resentment (Koglin, 2015). Lower-class suburban voters who are also car owners and sometimes reliable Social Democratic voters, resented the Copenhagen toll ring proposal (ibid.). In 2012 one suburban Copenhagen mayor claimed that "this system will make it easier for the rich to drive and force everyone else onto crowded trains" (Berlingske Media, 2012; City Finance Executive, 2016).

Gentrification would intensify because the cost of housing within the congestion zone, already inflated due to broader social trends such as reurbanization and the livable city, would increase further because of the added desirability to be inside the payment zone. This has arguably contributed to gentrification in London, although there are many other contributory factors (Short, 2018). While in Copenhagen redistribution of toll revenues might have looked promising, the toll ring, as with many mobility polices, did not take into consideration housing access or the broader right to the city concerns described in Chapter 4.

Class antagonism towards Copenhagen's toll ring may have been compounded by the global financial crisis, which despite the fact that it began in Denmark a little later than in the United States, did have an impact on car and home sales in Denmark from 2009 onwards. Copenhagen largely weathered the recession that dampened economic growth in the rest of Denmark, and so resentment in the hinterland was aimed at the city. As car sales rebounded in Denmark around 2012, resentment was bound with fears that the toll ring would discourage car ownership and punish those new post-recession car buyers.

Fairness has emerged with congestion pricing proposals elsewhere, especially in the failed New York City proposal of 2008 and again in 2018 (Short, 2018, Dwyer and Hujan, 2018). Yet in many cities where congestion pricing has been proposed, few low-income commuters travel by car into the proposed charging zone. Nevertheless, some mitigation of the cost burden on lower-middle-class motorists may have merit. Income-based discounts or waivers could mitigate equity concerns, giving tolling a Left/Progressive hue. However, in Copenhagen no income-based discount was proposed.

At least one local transport expert has suggested that the Copenhagen toll ring proposal should have included a graduated charge based on income, with wealthier motorists paying higher tolls and lower-income motorists paying substantially less (City Transport Executive (1), 2016). At the time of the toll ring proposal in 2011, an income-based approach was considered possible but difficult and unwieldy. With new technology, and the digital connection between license plates to personal and household income, the potential for a graduated toll is viable, but does touch on privacy issues. Arguments against the graduated pricing approach are easier to refute now because of new scanning and tracking technology, and other

Scandinavian countries have implemented income-based fines for speeding and other traffic violations.

The inability of Copenhagen's Left/Progressives to navigate the equity arguments are confounding. For example, the car system, as Harvey (2012) pointed out, is already deeply inequitable and especially for the poor and the working class. If implemented fairly, and with the goal of truly limiting cars and freeing more of the commons from private cars, proponents of the toll ring might have had a strong case from a left-leaning perspective (Freudendal-Pedersen, 2015). Copenhagen's debate, as was the case in both New York and San Francisco, seemed only to focus on the immediate impact that the toll would have on drivers worried about the individual cost to themselves, and not to the commons (Transport Scholar (2), 2016; Transport Consultant (3), 2016).

Conclusion: post-toll ring or congestion pricing renaissance?

The defeat of the toll ring was devastating for many of Copenhagen's green mobility professionals, politicians, and advocates (City Transport Planner (5), 2016). As one stakeholder and observer in Copenhagen's toll ring lamented, "Politicians don't want to address mobility truthfully" (ibid.). Sadly, this crosses the ideological spectrum, although Copenhagen's Red-Green Alliance has been the most insistent on car restraint. Some are resigned that nothing will be done about the car. During the debate about the toll ring, Copenhagen's Neoliberal business elite offered either tepid support for the toll ring, or remained silent. The Danish Confederation of Industry, which promoted Copenhagen's green branding, including the bicycle and wind power, notably made no mention of the toll ring or any kind of car restraint during its summits and in its publications (Transport Scholar (2), 2016). In *Sustainable Transportation Copenhagen* (2016), the State of Green organization omits any mention of the toll ring as a sustainable transport policy.

The business elite has acted similarly during other failed attempts to introduce tolling around the world. In San Francisco, the neoliberal San Francisco Planning and Urban Research, an erstwhile enthusiast of public transportation and cycling, was silent during San Francisco's congestion pricing debate in 2010. This reflects a broader trend of Neoliberals failing to organize around their principles of concern for climate change and urban livability when it comes to confronting the car.

Is it politically infeasible in Copenhagen to put limits on the car, by way of a toll ring or other approaches? With respect to the toll ring, many suggest that the 2011–2012 debate set congestion pricing in Copenhagen back at least a decade because there are bitter memories and deep ideological divisions that cannot be overcome until more people have a sense of urgency about the car (City Transport Executive (1), 2016). Others are hopeful that congestion pricing will come back in one form or another (City Finance Executive, 2016).

Beyond Copenhagen, there is a hint that 2019 or 2020 might give way to a congestion pricing renaissance as cities around the world confront the impact of new waves of cars and congestion. In early 2019, congestion pricing proposals were revived in New York and legislation in California was proposed to enable San Francisco to reconsider congestion pricing. Both proposals were still hemmed in by their respective state legislatures but pressure was mounting in New York State. In California in 2018 there was proposed legislation to enable San Francisco and a few other municipalities to pilot congestion pricing.

A new Progressive mayor of Seattle proposed congestion pricing in that city in 2018, and in Vancouver, Canada, planners initiated some tentative explorations. In London, new environmental "toxic charges" were deployed on the most polluting vehicles in 2017. This toxic charge zone covers the same geography as the toll ring, and targets vehicles registered before 2006, seeking to eliminate diesel vehicles and the dirtiest petrol vehicles. For these cars and trucks, the toll ring charge of $11.50 rises to £21.50 (i.e., up from US $14.00 to $30.00).

Other forms of car restraint are also germinating or expanding. In Germany, new restrictions on diesel cars were proposed in several city centers including Stuttgart (although not explicitly pricing). Environmental zones limiting cars are promoted by a handful of cities that are part of the C40 Cities network (C40 Cities, 2018). In Beijing, a lottery system is in place to meter car ownership in that city. In Mexico City, the "Hoy No Circula" ("Days off the Road") scheme restricts private cars based on license plate numbers, with the aim of 20 percent less car traffic every day (Cervero, 1998; Mahendra, 2008).

Santiago, Chile, uses a similar scheme to Mexico City, and also has an emerging new toll road network. São Paulo has a license-based car restraint scheme as well. Like car restraint policies everywhere, license-based restrictions have all resulted in heated political debate. More importantly, they are not making a substantive dent in traffic and pollution because of cheating the system because car owners are often purchasing a second car, or there is corruption in local policing. In Germany, too, there are questions about the efficacy of restricting cars in central cities if there is no real enforcement by police. All of this points back to congestion pricing and especially toll rings as strategies for making these cities livable.

Meanwhile, back in Copenhagen the leader of the municipal Social Democrats maintained that a toll ring is not politically feasible in Copenhagen because of national political opposition and because other road pricing methods might be preferred (Beim, 2018). For its part, the Red-Green Alliance continued to advocate for the toll ring. On the Right/Conservative spectrum, the Danish Road Directorate was once again suggesting a distance-based tolling scheme nationwide, in exchange for reducing the car tax.

In May 2018 there was renewed interest in congestion pricing at the municipal level with an adopted resolution asking that the city formally request enabling legislation from the national government (still led by the Liberals) and that the city prepare a new congestion pricing proposal.

Pushed forward by the new Alternative Party, the vote was 40–10 in favor of the resolution, with two members of the Danish People's Party abstaining (City of Copenhagen, 2018)

Voting in favor to pursue congestion pricing was the Left/Progressive majority on the city council: the Red-Green Alliance, the Socialist People's Party, and the Social Liberals. The Social Democrats and Social Liberals also voted for further study into the matter. Voting against were the Liberals, the Conservatives, and the Liberal Alliance (the Danish People's Party abstained from the vote). The Liberals accused the left of a back door attempt at imposing a toll ring, and suggested that air pollution could be reduced with the introduction of electric cars, and that the harbor tunnel would relieve congestion if it were built. The Liberals also called for an expansion of the Metro (which to the party represented more space for cars above the ground). The Liberal Alliance defended motorists, claiming that drivers were already stressed enough by the car tax and that it would never endorse congestion pricing unless the car tax was eliminated. On the right, the Liberal party and its allies continued in 2018 to insist on the distance-based option coupled with reduced car taxes.

The Alternative suggested openness to area-wide or distance-based tolling in lieu of a toll ring, if the evidence supported that option. This "intelligent road pricing" in the capital region could be part of the study objectives. Together with the Red-Green Alliance, the Alternative reminded the city that the climate plan could be realized without car restraint and that emissions must be reduced substantially more than even previously assumed. They added that in addition, noise, public safety, and the consumption of city space by cars were at stake.

The political feasibility and timing of the study has yet to be determined, but hope remains that Copenhagen may one day introduce either a toll ring or some other kind of congestion pricing scheme. Yet it is also obvious that the Right/Conservative bloc has doubled-down on the politics of the car, and has embraced technology as a solution rather than rethinking urban space.

Copenhageners, especially those on the left, are still asking: How many cars should there be in the city? In the meantime, there are steps which Copenhagen can take on the streets, and which have some potential. Copenhagen may have to wait many more years for the right politics of mobility in Denmark to become enabled to implement tolling, and Copenhagen cannot implement its own car tax, but Copenhagen does control the streets within the municipality, just as it controls the very cycle tracks that make it iconic. If few observers outside Denmark have been aware of Copenhagen's toll ring debate, it is likely that even fewer are aware that parking—both at the curb and in new developments—is often just as vitriolic and emotional.

References

Beim, J. H. 2018. "Mette Frederiksen: No to Payment Ring." *Politiken*, 22 August.
Berlingske Media. 2012. "Wild Battle for the Payment Ring." *B.T.*, 8 January.

Börjesson, M., Eliasson, J., Hugosson, M. B., and Brundell-Freij, K. 2012. "The Stockholm Congestion Charges: 5 Years On. Effects, Acceptability and Lessons Learnt." *Transport Policy*, 20: 1–12.

C40 Cities. 2018. *Fossil Fuel Free Streets Declaration.* New York, C40 Cities. Available at www.c40.org/other/fossil-fuel-free-streets-declaration (accessed 15 December 2018).

Cervero, R. 1998. *The Transit Metropolis: A Global Inquiry.* Washington, DC, Island Press.

Chronopoulos, T. 2012. "Congestion Pricing: The Political Viability of a Neoliberal Spatial Mobility Proposal in London, Stockholm, and New York City." *Urban Research and Practice*, 5: 187–208.

City of Copenhagen. 2009. *Impact of Copenhagen's Parking Strategy.* Copenhagen, Technical and Environmental Administration.

City of Copenhagen. 2012. *CPH 2025 Climate Plan: A Green, Smart, and Carbon Neutral City.* Copenhagen, Technical and Environmental Administration.

City of Copenhagen. 2013. *Action Plan for Green Mobility.* Copenhagen, Technical and Environment Committee.

City of Copenhagen. 2017. *Traffic in Copenhagen: Traffic Figures 2010–2014.* Copenhagen, Technology and Environment Administration.

City of Copenhagen. 2018. *Member Proposal on the Introduction of Road Pricing in Copenhagen.* Copenhagen, Copenhagen City Council. Available at www.kk.dk/indhold/borgerrepra esentationens-modemateriale/31052018/edoc-agenda/babf8d9b-aace-414a-999f-818a1a 5951fa/0bfbc6f5-ab91-4bf8-b9e3-34f51d75bc83 (accessed 30 May 2018).

City of Stockholm Traffic Administration. 2009. *Analysis of Traffic in Stockholm.* Stockholm, City of Stockholm Traffic Administration.

Dwyer, J. and Hujan, W. 2018. "Driving a Car in Manhattan Could Cost $11.52 under Congestion Plan." *New York Times*, 18 January.

Forum of Municipalities. 2008. *Congestion Charging in the Greater Copenhagen Area.* Copenhagen, Forum of Municipalities.

Freudendal-Pedersen, M. 2015. *Why Do We Need Utopias?* Available at http://en.forum viesmobiles.org/video/2015/05/22/why-do-we-need-utopias-malene-freudendal-peder sen-2873 (accessed 2 September 2016).

Gehl, J. 2010. *Cities for People.* Washington, DC, Island Press.

Hardin, G. 1968. "The Tragedy of the Commons." *Science*, 162: 1243–1248.

Harvey, D. 2012. *Rebel Cities: From the Right to the City to the Urban Revolution.* New York, Verso.

Homann-Jespersen, P., Sørensen, C. H., and Andersen, J. F. 2001. *Trafikpolitik og trafikplanlægning i Hovedstadsområdet (Traffic Policy and Traffic Planning in the Capital Region).* Roskilde, FLUX Center for Transport Research, Roskilde University.

Inrix. 2015. *Copenhagen Named Least Congested Scandinavian City.* Altrincham, Inrix. Available at http://inrix.com/press-releases/copenhagen-named-least-congested-scandinavian-city/ (accessed 14 November 2016).

King, D. A., Manville, M., and Shoup, D. 2007. "The Political Calculus of Congestion Pricing." *Transportation Policy*, 14: 111–123.

Koglin, T. 2015. "Organisation Does Matter: Planning for Cycling in Stockholm and Copenhagen." *Transport Policy*, 39: 55–62.

Mahendra, A. 2008. "Vehicle Reductions in Four Latin American Cities: Is Congestion Pricing Possible?" *Transport Reviews*, 28: 105–133.

Naess, P. 2006. *Urban Structure Matters: Residential Location, Car Dependence, and Travel Behavior.* London, Routledge.

Newman, P. and Kenworthy, J. 2015. *The End of Automobile Dependence: How Cities are Moving Beyond Car Based Planning.* Washington, DC, Island Press.

Oldenziel, R., Emanuel, M., de la Bruheze, A. A., and Veraart, F., eds. 2016. *Cycling Cities: The European Experience*. Eindhoven, Foundation for the History of Technology.

Rich, J. and Anker Nielsen, O. 2007. "A Socio-Economic Assessment of Proposed Road User Charging Schemes in Copenhagen." *Transport Policy*, 14: 330–345.

Ross, B. 2006. "Stuck in Traffic: Free-Market Theory Meets the Highway Lobby. "*Dissent*, 53(3): 60–64.

Short, J. R. 2018. "Are Traffic-Clogged US Cities Ready for Congestion Pricing?" *The Conversation*. Available at https://theconversation.com/are-traffic-clogged-us-cities-ready-for-congestion-pricing-90814 (accessed 26 December 2018).

Sørensen, C. H., Isaksson, K., Macmillen, J., Åkerman, J., and Kressler, F. 2014. "Strategies to Manage Barriers in Policy Formation and Implementation of Road Pricing Packages." *Transportation Research Part A: Policy and Practice*, 60: 40–52.

State of Green. 2016. *Sustainable Urban Transportation*. Copenhagen, State of Green.

United States Department of Transportation. 2011. *Congestion Pricing: A Primer*. Washington, DC, Federal Highway Administration.

Vickery, W. 1963. "Pricing and Resource Allocation in Transportation and Public Utilities." *Journal of the American Economic Association*, 53: 452–465.

Interviews held by Jason Henderson from 2015 to 2017 with transportation experts in Copenhagen

City Climate Planner (1), 2016.
City Finance Executive, 2016.
City Transport Planners (2 and 3), 2016.
City Transport Planner (5), 2016.
City Transport Planner (6), 2016.
Transport Advocate (2), 2016.
Transport Advocates, 2016.
Transport Advocate and Politician, 2015.
Transport Consultant (1), 2016.
Transport Consultant (3), 2016.
Transport Scholar (1), 2016.
Transport Scholar (2), 2016.

6

CYCLING POLICY IS PARKING POLICY

The politics of car parking in Copenhagen

Cycling policy is parking policy

In Copenhagen, as in most cities around the world, creating spaces for cycling requires trade-offs with the way in which street space is allocated. If there is plenty of space on a multi-lane arterial street or boulevard, the municipality might reduce the number of vehicle travel lanes, preserve on-street car parking, and place the cycle track between rows of parked cars and the sidewalk. When cycle track capacity is saturated, cycle tracks need to be widened to create passability for cyclists, and it might be necessary to remove the adjacent on-street car parking. Considerations for visibility at intersections and separating cyclists from buses means even more street space must be reallocated, and perhaps even more on-street parking removed.

On some of Copenhagen's narrower streets, and especially in more politically conservative Frederiksberg, cycle tracks feel more crowded because a decision was made to preserve on-street parking instead of widening the cycle tracks. On parts of Gammel Kongevej (Old King's Road), for example, cyclists are uncomfortably constricted as if in a cattle run, wedged between a narrow sidewalk and rows of parallel-parked cars. Here the city stencils warnings onto the sidewalk for cyclists to watch for car doors swinging open across the cycle track. The solution—widening the cycle track to provide room for a cyclist to avoid the door zone—would mean removing the adjacent car parking (see Figure 6.1).

Many of Copenhagen's neighborhood streets are jam-packed with parked cars and this makes cycling more challenging. Rows of parked cars impede cyclists' visibility, making it difficult for children to navigate safely, and cars, whether they are parallel-parked or parked at an angle, block visibility especially at intersections and corners. Throughout the city quiet neighborhood streets have angled parking with cars jutting into the street. While parking might be removed for a nearby

FIGURE 6.1 Narrow cycle track and parked cars on Gammel Kongevej in Frederiksberg.
Source: Photograph by J. Henderson.

cycle track or bus stop, parking on nearby streets has shifted to angled parking in an attempt to placate outrage by local car owners or merchants. This clutters and degrades what are supposed to be traffic-calmed and human-scale streets.

On narrower residential streets cyclists are allowed to travel in both directions but the narrow passage between curbs packed with parked cars can be intimidating as cars attempt to pass cyclists and force cyclists to steer awkwardly towards the parked cars. Cyclists in dense neighborhoods with bicycle parking deficits eye curbside street parking for more bicycle (and cargo bike) parking, bringing additional conflict between cyclists and motorists for curb space.

As with the toll ring, parking in Copenhagen is about how many cars should be in the city. With less car parking, there would be less room for cars, leading to fewer cars in Copenhagen, and more space that could be used for other purposes such as wider cycle tracks, pedestrian zones, or public transit space. Parking removal, increased parking pricing, and reduced parking in new housing areas are all types of proven car restraint policies and strong pathways to green mobility (Shoup, 2005; Foletta and Henderson, 2016; Buehler *et al.*, 2016; Kodukula et al., 2018). If more parking becomes available it is an invitation for more car ownership and easier use of cars (Shoup, 2005).

If Copenhagen allows more parking, by preserving on-street parking while building more off-street parking, the city's distinctive low rate of car ownership—

200 cars per 1,000 inhabitants—could vanish. If more parking is made available, Copenhagen may become more like peer cities in Europe which have much higher rates of car ownership, or worse, it could resemble traffic-clogged cities like San Francisco in the United States, which has a similar population density to that of Copenhagen, but which is stuffed with over 10,000 registered cars per square mile, thus undermining the city's reputation as a green mobility bellwether (Henderson, 2013; Foletta and Henderson, 2016).

The politics of parking matters and has an outsized role in what kind of city is created, and car parking is an especially vexing barrier to good cycling spaces. As one stakeholder, referring to what was characterized as the city's slow pace of widening cycle tracks, put it: "Copenhagen city cycling policy is parking policy" (Transport Advocate (2), 2016).

The spatial conflict between cycling and car parking is a volatile flashpoint in Copenhagen's politics of mobility. The Left/Progressives and pro-cycling parties such as the Social Liberals advocate for more cycling spaces and less car parking. The Right/Conservatives campaign against parking removal and demand more car parking. The Neoliberals claim to endorse market solutions for the allocation of space, but also oppose parking removal and demand more car parking, especially to market new upscale housing developments.

In addition to the battle for the curb, there is also the question of the price of parking, and the Left/Progressives, consistent with their support for the toll ring, and rather unlike Left/Progressives elsewhere, have endorsed much higher parking prices with the explicit goal of discouraging people from driving in Copenhagen. Contrary to their stated economic ideology, the Neoliberals balk at the introduction of higher prices even if they reflect the going market rate. Meanwhile, the Conservative Party insists that car owners should be free to park in the city (Conservative Party of Copenhagen, 2017).

At the national level, Denmark's Neoliberal-Right/Conservative green mobility coalition seeks to restrain Copenhagen's Left/Progressive inclinations towards using parking as a means of car restraint. In much the same way that the toll ring required that the Danish national government enable legislation, the Neoliberal-Right/Conservatives have claimed jurisdiction over Copenhagen's ability to increase curbside street parking prices. The national government also siphons off a hefty portion of Copenhagen's parking revenue so that Copenhagen cannot develop a redistributive parking policy by using parking revenue to fund green mobility. In all of this the politics of parking ultimately complicates efforts to realize Copenhagen's cycling and green mobility goals.

Copenhagen's Left/Progressive parking legacy

As pointed out by the Alternative party in the previous chapter, cars consume about 66 percent of Copenhagen's public right of way (City of Copenhagen, 2018a). They base this figure on analysis carried out by the city's Technical and Environmental Administration, which recognizes the spatial inequity of car parking on Copenhagen's streets (ibid., 2017c). While cyclists make almost 30 percent of all trips in and around Copenhagen, and 62 percent of all trips to places of work and education within Copenhagen, only 7 percent of Copenhagen's street space is dedicated to cycling (this does not include shared spaces on neighborhood streets). Cars, which account for 34 percent of all trips made in and around the city, but only for 9 percent of trips to places of work and education, have been allocated 54 percent of street space for movement, and 12 percent for street parking. More space in Copenhagen is dedicated to curbside car storage than to spaces for separated and safe cycling. To complete the picture, pedestrian space makes up 26 percent of road surfaces in Copenhagen.

It is this kind of spatial acumen that has informed Left/Progressive parking politics in Copenhagen. Copenhagen's Left/Progressives are willing to challenge the issue of car parking, and this is instructive to Left/Progressive politics elsewhere around the world where Left/Progressives overlook the profound impact that parking has on cities and equity. It is also instructive for Copenhageners who may have forgotten the legacy of the Left/Progressives in challenging the car through parking policy. Historically Copenhagen's Left/Progressives have supported the removal of parking as part of the reallocation of street space for green mobility.

In 1962 Copenhagen began restricting cars in the medieval core and pedestrianized the Stroget, which was, and remains, the flagship shopping street in Copenhagen. The scheme was rooted in a progressive, humble, social democratic appreciation for public space in a part of the city where it made no sense to encourage cars. At this time the city was crammed with cars, and it was decided that parking would be removed incrementally in order to avoid a major political backlash (Beatley, 2000). The goal, recognizing the spatial inequity and disproportionate impact of parked cars, was to "reconquer" space for pedestrians and halt the "invasion" of cars (Gehl and Gemzoe, 2003; O'Sulllivan, 2016). Curbside parking removal also helped to re-establish some cycle tracks in the 1980s.

Copenhagen's removal of street parking is legendary among some influential North American and English-language urban planners (as we will see, this was more mythology by 2018). Copenhagen's annual expropriation of curbside parking was lauded in *Transit Metropolis* (Cervero 1998). Beatley (2000) also refers to the annual reduction of parking since the 1960s, and notes that 600 parking spaces disappeared in a ten-year span. Newman *et al.*, (2009) and Newman and Kenworthy (2015) celebrated the fact that that for over 30 years, 2–3 percent of curbside parking had been removed annually in Copenhagen, and praised Jan Gehl for these street parking removal schemes. Locally, the inspirational narrative of street parking removal on the Stroget is highlighted in the business-oriented State of Green's 2016 *Sustainable Transportation Report* as a model of green urbanism. Copenhagen's parking removal strategy also inspired street parking removal and pedestrianization schemes in other cities around the world, notably New York in the 2000s (Sadik-Kahn and Solomonow, 2016).

To be sure, when the Stroget was first envisioned as a pedestrian street, parking politics was less contentious because Copenhageners and many Danes living outside the city owned fewer cars than they do today. The city was working class and was dominated by left-leaning politics. In the 1980s and 1990s, when cycle tracks were re-established and improved, Copenhagen was inhabited mainly by lower-income residents, including many students without cars, and so parking removal for cycle tracks was met with less political resistance than found today.

Parking removal was relatively cheap for city operations at a time when the city had fiscal problems. It was an inexpensive way to create public space or to lay out cycle tracks incrementally, while transforming the Indre By into a regional shopping destination and eventually a global tourist destination. Thus, to some extent Copenhagen's parking removal scheme had business inclinations. Many cities that implement public spaces through parking removal often do so with political compromise between the Left/Progressives and other business interests (Henderson, 2013).

Since the 1990s Copenhagen has had an extensive on-street parking payment zone that includes all of Indre By and the Brokvarterene (except Frederiksberg which is a separate municipality with its own parking payment scheme). Parking pricing, like the toll ring, would normally be associated with Neoliberal approaches to organizing space. Yet, in Copenhagen, the Left/Progressives have led on raising parking prices, because higher parking prices are a key part of green mobility if the goal is redistributive and coupled with improvements for cycling, walking, and public transport (Kodukula, 2018).

In 2006 citywide parking rates were raised by 50 percent by the Left/Progressives, when the Red-Green Alliance and the Social Liberals successfully pressured the Social Democrats to agree to this measure. In 2007 the payment zone was also expanded to cover most of the Brokvarterene. When on-street parking rates were increased and the geography of the payment areas expanded, car volumes in Copenhagen's urban core dropped by 6 percent, or by 18,000 cars at the various crossing-points over the lakes. Large employers in the city center reported that commuting by car to city-center offices dropped from 22 percent to 16 percent. High parking fees in the center (at US $5.55 per hour in 2017) encouraged people to cycle more and drive less in the city (Gössling, 2013).

Copenhagen's Left/Progressives consider the price of residential parking permits to be too low. These are special annual parking permits issued to car owners proving residency, and allowing the car to be parked for multiple days on designated streets. When the plan for Copenhagen's toll ring was scuttled in 2011, Copenhagen's Left/Progressives, including the Red-Green Alliance, the Socialist People's Party, and more moderate Social Liberals proposed that the price of an annual parking permit should be increased to 3,500 kroner (equivalent to US $542 in October 2018) (City Transport Planner (5), 2016). The Social Democrats, along with Copenhagen's right-wing parties, rejected that proposal (Transport Scholar (1), 2016). However, the price of residential parking permits in Copenhagen, which at the time cost less than $40 per annum at that time, were indeed raised. In 2016 it cost $5.55 per hour to park on the street in Indre By and along Copenhagen's lakes, and 730 kroner ($114) per annum for a resident to park on the street in Copenhagen's payment zones.

In 2018 the Left/Progressives, through the Technical and Environment Administration, proposed expansive parking reform and increased parking costs as a path towards achieving the Copenhagen 2025 Climate Plan's target for reducing car trips to 25 percent of all trips. Aligning parking policy to climate goals, criticizing the spatial inequity of cars, and proposing that pricing be redistributive, Copenhagen's *Parking Report 2018* bore the fingerprints of Left/Progressive politics of mobility.

The *Parking Report 2018* warned that private car ownership was increasing in the city, and that more private cars were flowing into the city from the suburbs, thus putting more pressure on the city to provide parking (City of Copenhagen, 2018b). It advised that providing parking invited more cars to come into the city, thereby undercutting green mobility goals such as the 25 percent target for car trips. The report revealed that residential parking permits outnumbered the available neighborhood street parking. Moreover, the report stated that the city simply did not have sufficient space to create public off-street parking, in order to replace parking spaces that had been removed from the streets.

The *2018 Parking Report* called for a regulatory simplification of city parking and for recommendations consistent with Left/Progressive politics of mobility in Copenhagen, including expanding the geography of the payment zones; increasing the price of parking; and reducing the time limits of street parking to shorter increments. Forcefully, the Left Progressives suggested that the cost of residential street parking permits should be increased to 10,000 kroner, or almost US $1,500 per annum, with the expectation that this would reduce the total number of cars seeking to park in Copenhagen by at least 6,000 and reduce car traffic by about 10 percent (thus freeing up more space to allocate for cycling and public transit) (Astrup, 2018; City of Copenhagen, 2018c). Upwards of 15,000 parking spaces would be "released" and potentially reallocated for other purposes. Deploying higher-priced variable pricing of on-street parking and expansion of the payment geography to all of outer Copenhagen would act as a congestion pricing mechanism discouraging cars from being brought into the city.

In mid-2018 the Red-Green Alliance, the Socialist People's Party, the Social Liberals, and the Alternative coalesced around the parking plan and moved that it be adopted in its entirety. Not surprisingly, Copenhagen's Neoliberal-Right/Conservative mobility coalition—the Social Democrats, the Liberals, the Liberal Alliance, the Conservative People's Party, and the Danish People's Party—voted to reject the proposal (see City of Copenhagen, 2018c). In 2018, and relative to the city's own mobility and climate goals and aspirations, it appeared that Copenhagen was stalling, or even falling behind, on the parking issue.

The right-wing backlash over parking in Copenhagen

Copenhagen's legacy of progressive street parking policies is in lockstep with efforts to reallocate street space away from cars and to promote green mobility and especially cycling. One important defining moment was the debate over reallocating

,the street space of Nørrebrogade, the busiest cycling street in Copenhagen, with upwards of 45,000 cyclists daily in 2016. Nørrebrogade, as we have described previously, is the spine of Copenhagen's densest neighborhood, Nørrebro, and is a vital commercial corridor. The ground floor of virtually every building is a neighborhood-serving business or is has been put to some kind of active use, with multi-story residential flats above. It incorporates classic late nineteenth-, early twentieth-century reform block housing and the archetypal Brokvarterene, with no off-street parking. Copenhagen's most popular bus line, the 5A, is routed along Nørrebrogade as it traverses Copenhagen.

In the early 2000s Nørrebrogade, then four lanes with intermittent on-street parking, was congested with car traffic, parked cars, delivery vehicles, buses, constricted sidewalks, and narrower cycle tracks that were far from capable of handling what was then roughly 30,000 cyclists daily. It was a classic mobility stalemate whereby every user in every mode was dissatisfied if not miserable, and the only path forward was rethinking the street.

For Copenhagen's Left/Progressives, Nørrebrogade was also a symbol of the spatial inequity of the car. Per capita car ownership in Nørrebro was (and still is) the lowest in all of the Copenhagen region, at 115 cars per 1,000 persons, and Nørrebro was the only district in Copenhagen where car ownership rates were actually declining, despite increased population and higher incomes (City of Copenhagen, 2016). Clearly, too much space was being devoted to the car, while cyclists and bus passengers were impeded. The logical solution was to remove cars, and improve the flows of cyclists and buses.

Initiated by the Social Liberals, whose leader, Klaus Bondam, headed the Technical and Environment Administration between 2005–2009, Nørrebrogade was studied, redesigned, debated, and finally traffic lanes and cycle tracks were reallocated. Three proposals were put forward. The most far-reaching of the three, which had some support from the Left/Progressives (and including the Social Liberals), was a scenario that completely banned private cars from most of Nørrebrogade, creating center-running bus-only lanes with extra-wide cycle tracks straddling the busway (Transport Consultant (3), 2016). Some observers also suggested doing away with the cycle tracks altogether, and simply allocating the entire street to cyclists and buses in a shared roadway (with wider sidewalks on either side of the street).

Predictably, the Right/Conservatives and a group of merchants in the Nørrebrogade corridor objected to the reallocation and car restraint (City Transport Planner (4), 2016). A simplistic media sensation (resembling hysteria over congestion pricing) ensued, including a narrative that making more space for cycling and taking away car space would cause shops to go out of business. The narrative was easily refuted by the city. As described earlier, for example, 32 percent of trips to supermarkets and street-level shops in Copenhagen were by bicycle (City of Copenhagen, 2015a). An equal number of shopping trips were by car, and 23 percent of all citywide shopping trips were by walking. These are citywide figures. The city argued that in Nørrebro the rate of walking and cycling to shops far surpassed driving.

At this time (2008–2010) Copenhagen merchants were feeling the impact of the global financial crisis. While perhaps rightfully jittery, major exogenous factors with the economy were affecting small businesses, not just in Copenhagen, but worldwide. Meanwhile, Nørrebro was becoming gentrified, and this included commercial gentrification. Rather than focus on financial support for small businesses, or addressing commercial rent inflation, the easiest target, for Right/Conservative factions, was to blame bicycles for apparently not shopping, and to claim that car access was necessary for local businesses to stay afloat (Saehl, 2013).

The narrative of economic decline due to cycling got a lot of press, but it was not potent enough because the evidence was on the side of cyclist (and bus passengers) (Saehl, 2013). Seven city departments, meaning a substantial majority of Copenhagen's governing structure, endorsed the proposal to make Nørrebrogade car-free. This, it should be noted, was unprecedented up to that date, because Copenhagen's city departments were notorious for not being able to work together (Transport Consultant, 2015; Transport Advocate and Politician (1), 2016).

Yet despite near unanimous city support, the Danish police, which as we have suggested leans Right/Conservative on Copenhagen street politics, vetoed the car-free proposal. Recall that the police, which are under the jurisdiction of the Ministry of Justice, can veto any parking or traffic change that is deemed "unsafe." In the case of Nørrebrogade, police leadership argued that parking removal and restrictions of cars on Nørrebrogade would increase pressure for parking on adjacent streets, and divert traffic to adjacent streets. All things considered, this was probably true in the immediate short term. However, over time, and with a multipronged approach of enticing drivers out of cars through parking management and congestion pricing, coupled with bicycle and transit improvements (the Metro City Ring Line was about to commence construction), the initial traffic impacts would settle down and traffic would adjust.

The Danish police disagreed and instead embraced a Conservative essentialization of the car. With or without parking, cars would come to Nørrebrogade and cars needed to come to Nørrebrogade. Using arguments made by police about public safety due to increased traffic caused by parking removal and restrictions on cars, the Right/Conservatives had found their technicality without needing a democratically elected majority on the Copenhagen City Council. The car-free Nørrebrogade proposal, like congestion pricing and other ambitious green mobility projects that circulated in Copenhagen at this time, was scuttled.

The Left/Progressives were set back but determined to transform Nørrebrogade. An alternative scenario that limited but did not ban cars was offered and accepted by the City Council and the Danish police backed down. This scenario removed on-street parking on most of Nørrebrogade and addressed concerns about merchant deliveries. Instead of banning cars for the length of the roadway, several forced right turns were installed at key intersections. Special bus-only lanes with red sidewalk markings redefined certain spaces where cars were forced to turn right off of the street.

With forced right turns, cars could navigate the street in short segments, but would then have to deviate off of Nørrebrogade. Nørrebrogade would no longer be a principal through road for driving. Nørrebrogade's cycle track was widened to Plus-Net level to accommodate passability for cyclists riding three abreast. This improved conditions for cycling tremendously, but the sidewalks were not widened. Therefore, pedestrians remained on constricted sidewalks, which were often cluttered with fly-parked bicycles, because adequate bicycle parking was, and still is, lacking in this part of Copenhagen.

Nonetheless, the 5A bus sped up. Cycling increased from 30,000 to over 45,000 cyclists per day. Today Nørrebrogade also throngs with pedestrians even though the sidewalk is constricted and cluttered. There are few storefront vacancies and Nørrebrogade is one of the most desirable neighborhoods in which to live and to visit. And on weekday mornings, pelotons of cyclists commute in near silence interrupted only by passing buses or occasional delivery trucks. The absence of revving engines and particulate pollution from car traffic is very noticeable when compared to other roadways like H. C. Andersens Boulevard.

The Nørrebrogade compromise was adopted by a two-thirds' majority in the Copenhagen City Council (Transport Advocate and Politician (1), 2016). Tellingly, despite the compromise, it should be noted that one-third of the City Council was still skeptical about parking removal and car restraint, even on the city's busiest cycling street, with the city's highest bus ridership, the city's highest population density, and lowest rates of car ownership.

This right-wing objection, coupled with media sensation about angry motorists and victimized merchants, had a long-lasting chilling effect on Copenhagen's green mobility ambitions, and is inextricably bound with the demise of congestion pricing and the weak mobility policies in the city's climate plan. Once implemented, the media turned its attention to car drivers who were angry about the forced right turns on Nørrebrogade, and the Social Liberals were especially punished at the polls. Klaus Bondam, who at the time was city mayor and head of the Technical and Environment committee, described the fallout as "Bondam bashing" by irate motorists sensationalized in local media (Transport Advocate and Politician (1), 2016). In 2016 he claimed that he was still receiving hate mail for the Nørrebrogade episode.

When street reallocation was first proposed for Nørrebrogade the city had an ambitious bicycle planning agenda and had adopted the Plus-Net scheme of prioritizing 3-meter-wide cycle tracks on main arterials, while also eliminating bicycle/bus conflicts by introducing reconfigured bus stops and routing cycle tracks around bus stop islands (examples of which were part of the Nørrebro compromise). This template was to be repeated on a handful of other arterial streets in Copenhagen, such as Vesterbrogade and Amagerbrogade, but the negative effect of a Right/Conservative backlash stifled expansion of the Nørrebrogade template even though a compromise was eventually found.

Since the Nørrebrogade episode, a handful of other bicycle and bus priority plans have either been shelved or significantly watered down (Transport

Consultant, 2015; City Transport Executive (1), 2016; Transport Consultant (2), 2016; Transport Consultant (3), 2016; City Transport Planners (2 and 3), 2016). The political will to remove car parking and reallocate streets seemed to have diminished, if not exhausted.

In Vesterbro, twenty-years of plans for cycle tracks on Istedgade, a neighborhood-commercial street, were dropped in 2015 following merchant opposition (City of Copenhagen, 2015b). To address complaints about the parking removal, parking on neighboring residential streets was shifted to angled parking, and on-street parking was also preserved. As with Nørrebrogade, there was compromise, and rather than cycle tracks, the city widened the sidewalks at the behest of merchants. Cars continue to speed along the street from time to time, and near-brushes with cyclists occur (City Transport Planner (1) 2016; City Transport Planner (6), 2016).

On Vesterbrogade, just a few meters north of Istedgade and the main street of Vesterbro, the first narrow cycle track was built in 1993. When the city considered widening cycle tracks for capacity improvements, on-street parking was a key debate. The compromise was a 'road diet,' referring to the conversion of four travel lanes into two with periodic center-turn pockets. A cobblestone median was placed in the middle of the roadway, making it difficult for cyclists to turn left if their destination is mid-block but across the street. The scheme actually added capacity for cars by 25 percent, and put a steady flow of cars on the street. Narrow car lanes do force slower, more attentive driving. Vesterbrogade does not have forced right turns like Nørrebrogade, and so remains a through road. It carries over 42,000 cyclists per day, up from 36,000 in 2012, but Vesterbrogade also carries more than 30,000 cars, buses, and trucks (Transport Consultant, 2015; Transport Consultant (2), 2016; Transport Consultant (3), 2016; Transport Scholar (1), 2016).

In Amager, Amagerbrogade came under considerable debate between 2012 and 2016, and there was dissatisfaction among the Left/Progressives about the outcome. In a plan similar to that implemented on Nørrebrogade, cycle tracks were widened, but to only 2.8 meters, not to the 3-meter Plus-Net standard. Proposed extended bus lanes were dropped although bus improvements were made, including signal optimization. Short-term on-street parking was also preserved along the outer reaches of Amagerbrogade, further from the city center (City Transport Planner (4), 2016; City Transport Planner (6), 2016; Transport Scholar (1), 2016).

In Østerbro, the configuration of Nordre Frihavnsgade has been debated many years. Like Nørrebrogade and Istedgade, it is a main shopping street, and so there is strong merchant opposition to parking removal in order to install cycle tracks. Østerbro is the wealthiest part of the Brokvarterene and one of the wealthiest districts in Copenhagen. The shops are higher-scale boutiques, and shopowners insist that parking is a necessity. Instead of cycle tracks, in 2018 the city was considering traffic calming and a bicycle boulevard treatment, yet the narrow street remains awkward for cyclists to comfortably circumnavigate given the moving cars on the left and parked cars on the right (City Transport Planner (1), 2016; Transport Consultant (2), 2016). In early 2019 plans for Nordre Frihavnsgade were delayed again (Astrup, 2019).

There are other examples of political recoil and downgrading of cycling proposals in the city. The political decisions about street parking have contributed to a mobility stalemate in Copenhagen whereby cycling infrastructure cannot expand, but car spaces and parking are preserved.

Parking "agreements"

The historic progressive narrative of incremental parking removal, despite the legendary accolades by sustainable transport scholars and planners, was by 2015 inaccurate and over-romanticized. That year several dozen curbside parking spaces were removed in the Indre By, but almost 100 added (through shifting to angled parking) in the Brokvarterene (City of Copenhagen, 2016). By 2016 the official city parking policy, with the slogan "room for everyone" was to identify 4,000 new parking spaces in a clear effort to accommodate the rise in the number of new cars in the city (ibid.; City Transport Planners (2 and 3); 2016).

The back-pedaling on parking removal reflected periodic, politically negotiated "parking agreements" made since 2005. Copenhagen's parking agreements require that any proposed removal of on-street parking be counter-balanced with a one-to-one replacement. The replacement parking can come from newly constructed, publicly accessible parking garages, on the condition that the garage is proximate to where the parking was removed. Additionally, one-to-one replacement parking can be created by shifting parallel curb parking to angled parking (Transport Consultant (3), 2016). Between 2005 and 2012 about 2,000 new on-street parking spaces were created in inner-city Copenhagen by shifting parking this way (City of Copenhagen, 2016).

The parking agreement slows the pace of reallocating street space. The city has not shown the building of new public garages in neighborhoods to be feasible. Lack of land, and very high land costs, coupled with the limits on how much more space can be captured for angled parking, means one-to-one replacement is increasingly difficult. This has hampered cycle planning and green mobility in Copenhagen. Since 2008 only three new parking structures have been built as part of the parking agreement, amounting to about 840 parking spaces. All of them cost more than originally expected, with significant cost overruns. The city also purchased two private garages with a total of 300 parking spaces. The new off-street parking spaces are available to any Copenhagen motorist holding a residential parking permit, and require payment for any non-residential permit holder. As with on-street parking there is no reserved or guaranteed available parking.

By 2009 the city was revising downward the number of new off-street parking spaces that were considered feasible to build and so also slowed parking removal. Only 200 on-street spaces had been removed, mostly in the Nørreport Station area and for the optimization of bus lanes near the Botanical Garden (City of Copenhagen, 2009). The city dropped the plan to remove an additional 1,000 parking spaces, thus stifling green mobility planning.

The "Parking Experience," a 500-car underground garage at Sankt Annæ Plads (St. Anne's Place), just north of the medieval core and Nyhavn, and adjacent to the

Royal Danish Playhouse, exemplifies how expensive replacement parking can be. After Copenhagen experienced rainfall-induced flooding in summer 2011, the city embarked on a climate and cloudburst adaptation strategy to create permeable surfaces for stormwater run-off. Approximately 500 on-street parking spaces were removed in the vicinity to create permeable semi-sidewalk and park-like green spaces for stormwater drainage. Nearby, 500 parking spaces were slotted into a new underground parking structure adjacent to the Playhouse on the harbor. Improvements for cycling were also included in the scheme, although stormwater management was the primary objective for the parking removal (City Transport Planners (2 and 3), 2016).

The St. Anne's Place parking scheme was controversial and expensive (700 million kroner, or US $118 million in 2008, for the garage, corollary streetscapes, park, and stormwater management). Some local car owners objected to the scheme because they did not get a discount to park in the new publicly accessible but privately operated garage. From the Left/Progressive side, there was criticism that keeping so much parking in the center of the city undercut the city's green mobility agenda, and that the new parking should, if really necessary, have been constructed on the outskirts of the city to intercept cars before they entered the urban core (City Transport Executive (1), 2016). Critics also pointed out that there were several public and private garages, such as Kongens Nytorv in the medieval core, and Israels Plads in the Nørreport section of Indre By, that were never full and which could be used to offset the removal of on-street parking (Transport Advocate and Politician (1), 2016).

In 2015 Copenhagen's revised parking agreement solidified the Neoliberal-Right/Conservative insistence on a one-to-one replacement. In that agreement, the Social Democrats made a deal with the small Neoliberal and Right/Conservative parties at the City Council to allow more parking in the city center and in new residential developments in the harbor districts (Transport Advocate and Politician (1), 2016). In a blow to the Red-Green Alliance, the Social Democrats also agreed with the Right/Conservatives that Copenhagen would no longer study or discuss a Left/Progressive proposal for a car-free medieval core (we will return to this issue in Chapter 7) (City Climate Planner (1), 2016).

Cycling policy is parking policy. If it is deemed too expensive to build one-to-one replacement street parking for cars, or impossible to shift to angled parking, then street parking cannot be removed as part of cycling improvements. The mandate of one-to-one replacement has effectively hampered the expansion of green mobility in Copenhagen, despite the city's reputation for incremental parking removal and a broader green mobility image.

The politics of parking pricing

As difficult as it has been for the Left/Progressives to remove parking, Copenhagen's politics of mobility also ensures that increasing the cost of parking in Copenhagen remains similarly difficult. To temper Copenhagen's Left/Progressive inclination to dramatically increase the cost of parking, the national (right-wing) government directly intervenes by collecting the parking revenue, siphoning off a proportion, and returning the balance to Copenhagen, ostensibly because the national government views parking costs as a tax, which it alone has the authority to impose (Transport Consultant (3), 2016).

Balking at some of the Left/Progressive proposals to increase residential permit parking on streets, in 2016 the national government characterized parking as a form of congestion charging, which as we have seen, is illegal without national enabling legislation (City Transport Planners (2 and 3), 2016). Moreover, the city can only petition to charge for parking if it can prove that this measure will decrease congestion. As we have also seen, the national police believes that parking removal might lead to congestion. High-priced parking, the logic follows, might also increase car circulation by motorists in search of cheaper parking.

By 2017 the city increased the cost of the residential parking permit but the Social Democrats and Right/Conservative factions were digging in to prevent any further increases. Providing the Social Democrats with political cover, in 2018 the national Neoliberal-Right/Conservative coalition government began to interfere in Copenhagen's parking politics. Led by the Liberals and the Liberal Alliance, as well as the Conservative People's party and the Danish People's Party, the national government adopted legislation that redirected 70 percent of the revenue collected from parking in Copenhagen to the treasury instead of to Copenhagen. At the time, 40 percent of revenue was already funneled to the national police (City of Copenhagen, 2015b).

The national Right/Conservatives also proposed to cap Copenhagen's ability to increase parking prices (recall that the Left/Progressives were proposing to introduce annual residential permits costing upwards of US $1,500). Some of the rhetoric used by the right charged that Copenhagen was attempting to use parking as a "cash machine" and a "municipal money harvest" from car drivers (Christiansen, 2018).

The national government argued that it was justified in implementing this scheme because by law municipalities were only legally allowed to raise parking revenue to make parking management schemes self-financed, but not to raise surplus money for other non-parking management purposes. Copenhagen's Left/Progressives, the Neoliberal-Right/Conservative coalition claimed, were bending the rules and using increased parking pricing as a cycle track expansion policy and other transportation improvements, which, they claimed, were city services that were not directly part of parking management.

In other words, for the Neoliberals and the Right/Conservatives, parking reforms should not be deployed as redistributive policies. Consistent with this attack on progressive parking, the Liberal Alliance claimed that only 30 percent of what Copenhagen collected was needed for the parking management scheme, and that the other 70 percent was simply raising revenue for the local government (Astrup, 2018). In late 2018 the Ministry of Economic Affairs and the Interior skimmed 200 million kroner annually from Copenhagen's parking meters (US

$30.5 million in October 2018). Had it remained in the city this money could have funded a significant number of cycle tracks or other local transport improvements.

During the 2019 Copenhagen budget debate (finalized in September 2018) the rift between the Left/Progressives and the Social Democrats widened over parking and other aspects of the politics of mobility including finance for cycling and transit and affordable housing. The Red-Green Alliance and the Alternative protested loudly and accused the Social Democrats of being bought off by the national right wing (Enhedslisten, 2018) (the Left Progressives were also factionalizing, as the Socialist People's Party and the Social Liberals signed the budget agreement with the Social Democrats). The Red-Green Alliance highlighted that it proposed higher parking fees that should be directed at cycle tracks, transit, and affordable housing, but that the Social Democrats, agreeing with the right wing, endorsed lower fees, reduced dedicated funding for cycling and housing, and shifted some city revenue to building off-street parking (Pedersen, 2018).

The Alternative's leader went as far as to state that "you [the Social Democrats and the party's allies] have chosen to lower the parking fee so it will be cheaper to park in the city. We think this is a stupid priority when we could invest more in cycling culture—in cycle superhighways and safer school routes for our children" (Pedersen, 2018). The 2019 budget included 75 million kroner for cycling (US $11.6 million in 2018), and both the Alternative and the Red-Green Alliance argued that the funding was insufficient because the Social Democrats and the right wing refused to increase parking prices. It also reflected that they wanted to slow the expansion of cycling infrastructure, because many projects had to be delayed due to funding shortfalls.

This appeared to be a Neoliberal-Right/Conservative victory on slowing green mobility. Furthermore, reflecting the Social Democrats' pro-car position, Copenhagen identified 30 million kroner (US $4.6 million in 2018) for initiating construction of new parking structures in Indre By and the Brokvarterene. As we have already seen, though, it is difficult to identify suitable space, and very expensive to construct new garages in dense neighborhoods. Parking politics in 2018 further stifled Copenhagen's cycling policies.

Parking in the new Copenhagen

In the Brokvarterene, Indre By, and many of the older sections of Copenhagen, the politics of street parking is a high-profile flashpoint in the struggle over green mobility. Debates about street parking are emotional and tangible, with car owners, cyclists, pedestrians, and transit passengers all jockeying for political control over the same curb space. Meanwhile, away from the street, behind or beneath buildings, less visible and less tangible private, residential off-street parking policies are an equally important flashpoint in Copenhagen's politics of mobility. Sometimes characterized as below the radar by local observers in Copenhagen, residential and commercial off-street parking has a profound and long-reaching impact on what kind of city is built and who that city is for.

The literature on the impact of off-street parking is vast and empirically unequivocal (Shoup, 2005, synthesized much of it). Off-street parking attracts cars, generates more car trips, consumes vast amounts of urban space that could be used for other purposes, degrades the pedestrian realm, privatizes portions of the curb where driveways intersect with streets, and adds a considerable increase to the construction costs of new housing, which is then bundled into higher rents or sale prices for apartments and housing. This last point is important for the broader housing affordability issue. In Copenhagen, an average flat is 10 percent cheaper without parking than with parking (Transport Advocate and Politician (2), 2016). For these reasons cities around the world are undergoing lively debates about what constitutes the appropriate amount of off-street parking, and Copenhagen is no different in this regard. Yet one would not know that off-street parking was a political flashpoint if relying on the boosterish pamphleteering that celebrates Copenhagen as a green mobility capital.

Copenhagen's canon of sustainability reports and branding materials all but ignore off-street parking. For example, parking is almost entirely ignored in Copenhagen's 2025 Climate Plan. The mobility sections of the climate plan include extensive promotion of improving cycling and public transit but, just as it omits the uncomfortable debates over street parking, it furthermore fails to acknowledge that off-street parking provision in new housing or offices undercuts building energy efficiency (the climate plan does include a paltry proposal for on-street car-sharing pilot programs).

Boasting of the new housing and commercial developments in the Ørestad redevelopment district on Amager, the *Copenhagen Climate Solutions Report* completely ignores the issue of off-street parking (City of Copenhagen and State of Green, 2014). The subsequent *2015 Green Account for Copenhagen* acknowledges the need to expand cycling capacity in the city but fails to mention how adding more parking in the city's new redevelopment areas will generate more car trips, and thus increase demand for the limited amount of road space that is already contested (City of Copenhagen, 2015d). The State of Green's *Sustainable Transport Report* (2016) also provides ample discussion about prioritizing cycling and pedestrians but does not offer suggestions to the city on what an appropriately sustainable amount of off-street parking should be.

Reports stemming from outside Denmark that celebrate Copenhagen as a green mobility capital also lack discussions about off-street parking. The London School of Economics (2014) barely mentions parking, suggesting only that some kind of vague regulation might be needed. The Brookings Institute, based in the United States, cheered the massive harbor redevelopment schemes, coining them the "Copenhagen model" of urban regeneration, but failed to mentioned the new parking that is accompanying new housing and commercial buildings (Katz and Noring, 2016). The Danish Architecture Center's *Guide to New Architecture in Copenhagen* (2015) says almost nothing about parking despite parking having a substantial impact on how buildings are planned, constructed, and built. In the case of off-street parking, the silence is deafening, and also very curious. It is therefore necessary and important to dig beneath the superficial veneer of green mobility promotion, and take a hard look at Copenhagen's off-street parking politics.

In cities around the world, the terms parking "minimums" and parking "maximums" commonly refer to floors and ceilings for the amount of parking required in new developments (older existing developments that predate off-street parking regulations are normally grandfathered unless substantially renovated or expanded). Parking minimums refer to the required amount of parking that must be provided in new developments. For example, in some of Copenhagen's redevelopment districts the minimum is one parking space for each 200 square meters of living space, expressed as a ratio of 1:200 m^2 (in the United States the ratio is more often based on the number of parking spaces per housing unit, such as 1:1, or one parking space for each new housing unit). Parking maximums, or caps on the allowed amount of parking, are similarly indicated as a ratio, and throughout the city of Copenhagen the parking maximum for new residences is 1:100 m^2.

Requirements for parking are a municipal affair in Denmark, and in Copenhagen, which adopted regulations in the 1950s, minimums and maximums are outlined in the city's Municipal Plan, which is in the bailiwick of the City's Finance Administration rather than the Technical and Environment Administration. This means that, while the Technical and Environment Administration pushes for more Left/Progressive parking policies on the city's streets, parking for private off-street parking is shaped by the Lord Mayor, a position that has been held by a Social Democratic incumbent for decades. We will come back to this shortly, but for now it is important to consider that Copenhagen could mandate a zero minimum if the city chose to do so, which would allow new residential or office developments the option to exclude parking and encourage car-free development. Childcare, fitness centers, community rooms, indoor pools, or community gardens could take the place of parking. In effect, this would be replicating the urban development pattern of the Brokvarterene, which has a tout ensemble that enables people to live comfortably and car free, while organizing a lifestyle around cycling.

Copenhagen presently does not allow new car-free development, even though zero minimums are preferable for green mobility and are widely recognized as a key step towards making cities less car dependent. Instead, Copenhagen, despite its celebrated status as a green mobility city, stipulates residential off-street parking with a minimum of 1:100 m^2 citywide, with exceptions in some redevelopment areas (described below). Copenhagen's citywide minimum and maximum are the same ratio, meaning that in effect each new housing development in Copenhagen is required to provide no less and no more than one parking space per 100 square meters of living space.

Historically the lack of private off-street parking has been fundamental to what made Copenhagen a green mobility capital. Even though Copenhagen adopted parking minimums and maximums in the 1950s, decades of economic decline and decreasing population meant that little new housing was built in the city until the 1990s and so off-street parking may have been less of a flashpoint compared to today (City Transport Planners (2 and 3), 2016). Today, with massive new redevelopment and tens of thousands of new residences either proposed or already under construction in the harbor and Ørestad, off-street parking requirements will undermine and complicate Copenhagen's aspirations to remain a green mobility capital.

In the late 1980s and 1990s, when Copenhagen was caught in a spiral of deindustrialization (see Chapter 2 in this volume), the city and national government, joint owners of former industrial and military lands, partnered to attract higher-wage, wealthier households to the city, which in turn would generate tax revenue (Andersen and Jørgensen, 1995; Hansen *et al.*, 2001). Because the Neoliberal-Conservative national government objected to raising taxes in order to underwrite urban regeneration in Copenhagen, and sought market-based solutions, the scheme was that the municipal and national government would clear and prepare the lands under the guise of a land development authority, while a separate entity would construct new Metro lines that would connect these redevelopment zones to the city center and to the airport.

The redevelopment scheme had a neoliberal hue in that it required subdivision of the land in the former port and military bases and disposal to private developers for a profit. Land sales to the private sector would be sequenced in a way that maximized profit, and took place only when the market appeared to be at its highest. The land sales must make profit for the city, and it was also assumed that Metro construction would increase land values. The profits would pay off the debt engendered by the Metro construction, as well as help to finance future Metro construction. Revenue from the land sales could not be redirected to other purposes, locking in the Metro as Copenhagen's preferred public transport for decades. Yet parking policy would also lock in the car for decades.

At first, when parking requirements were debated for the first stages of the Ørestad plan, the outcome was more progressive than some more recent parking politics in Copenhagen (reflecting a more left-leaning Social Democratic outlook in the 1990s compared to today). Planned as a model for sustainability, with heathland open space protection and especially climate mitigation, the Ørestad plan included off-street parking minimums at 1:200 m^2, meaning roughly one parking space for every two or three units of housing depending on the size of units (on average in 2014 a typical new flat build in Copenhagen had a floor area of about 80.5 square meters). Significantly, many of the new blocks of flats in Ørestad were proposed with no parking within buildings, and instead off-street parking would be allotted to separate, stand-alone, centrally located parking structures.

Decoupled from the cost of housing, in Ørestad residents with a car were obliged to rent a parking space. Today, with several large parking structures built amid clusters of flats, the price of parking is roughly 1,000 kroner (US $150 in 2017) per month. This rate is over ten times the cost of an annual residential on-street parking permit in the rest of the city—including in the Brokvarterene, and more in line with Left/Progressive policies described in the previous section. Additionally, much of Ørestad was built with no on-street parking; instead, cycle tracks and sidewalks straddle the curb. Where on-street parking is available, it is limited to two hours for visitors coming by car.

Ørestad's population in 2016 approached about 10,000 and the final build-out includes 25,000 residents, 20,000 students at new university campuses, and 60,000 jobs within the 1.2 square mile district. The first office buildings were completed in 2001 and by 2016 there were 17,000 jobs in Ørestad. Both the M1 and M2 Metro lines have also been completed. Yet since the initial more progressive parking policies of the 1990s, off-street parking has proliferated in Ørestad. Most noticeably, large amounts of surface parking were built adjacent to the Metro station at the Danish Radio Music Hall Station, and there is a massive amount of parking surrounding the Bella Center, one of the largest congress halls in Scandinavia. About 3,000 parking spaces are situated beneath the Field's shopping mall, and in 2018 several new multistory parking structures were under construction next to new residences, suggesting that car ownership and car use were on the rise in Ørestad, despite all the other green accolades that it has garnered.

During the initial stages of Copenhagen's harbor redevelopment along the central section of the inner harbor (straddling Indre By), large banks, offices, and hotels provided abundant underground car parking for office workers. Ironically, a media office building located directly between the touchdown for the Snake Bridge and the Bryggebroen crossing, both of which are major cycling corridors, is littered with surface car parking adjacent to thousands of passing cyclists, and occasional conflict can arise. At the other end of the Snake Bridge, the Fisketorvet Mall, or Copenhagen Mall, with thousands of parking spaces, boasts that "parking should be easy" and offers two hours' free parking to shoppers.

A stone's throw from a major S-train station and adjacent to high volumes of cyclists crossing the harbor via the Snake Bridge and the Bryggebroen, parking occupies about one-third of the total area of Copenhagen Mall, including the rooftop. To be fair, roughly 20 to 25 percent of customers of the mall arrive by bicycle using 650 bicycle parking spaces in a prominent bicycle parking facility at the main entrance. However, the access, especially from the south, is decidedly car oriented. Meanwhile, the area around the Mall and along the Kalvebod Brygge (Ring 2) is a drab, car-oriented environment even though it is located just to the east of the main railway station and Tivoli Gardens.

As building in Ørestad was underway and the inner harbor sections filled in with car-oriented offices and commercial activity, next came redevelopment of Sydhavnen (South Harbor), which was one of the largest residential redevelopment projects in Scandinavia when vetted in 2000 and at a time when parking provision became a key element to attract new wealthier households to Copenhagen (Hansen et al., 2001). Unlike Ørestad, Sydhavnen was built under the citywide parking ratio of 1:100 m^2, two times the minimum in Ørestad, despite being physically closer to the city center. This indicates a regressive and more politically conservative departure from the Ørestad development planned just a few years earlier.

With 15,000 residents expected on completion in the early 2020s, Sydhavnen was awkwardly constructed in phases beginning on the south side of the harbor and incrementally built northward towards the Copenhagen Mall. According to the original plan, the sequence of construction from south to north made sense because the city planned a new S-train station at the southern edge of the redevelopment area, furthest from the city center. This "infill" station would have been established on the mainline DSB electric railway between Copenhagen Central Station, the international airport, and the Øresund crossing to Sweden, which would have made Sydhavnen an ideal regional-scale public transit-oriented development. Unfortunately the station was never built, due in part to funding issues and in part to the concern of the Danish Railways that it would slow the higher-speed intercity service to Sweden (National Transport Planner, 2017).

Despite the unbuilt railway station, which arguably was the lynchpin for green mobility at Sydhavnen, development was allowed to proceed at the southern end in what is called Slusenholmen (Sluice Island). Yet instead of a new railway station, a network of fast, car-oriented arterials now border Sydhavnen, connecting the regional motorway network to the city center of Copenhagen and Ring 2. Sydhavnen resembles a high-density, car-oriented development (rather like Los Angeles) instead of a traditional Copenhagen bicycle and public transit-oriented development.

With abundant parking both underground and on the surface, Sluseholmen is sometimes derisively called "parking with the most beautiful view" by some sustainable transport advocates (Transport Advocate (3), 2016). Completed in stages between 2006 and 2008 Sluseholmen has flats with underground parking and direct elevator access between flats and the parking, making is much easier to access a car than in Ørestad. The price of parking is also substantially less than in Ørestad, at 550 kroner per month instead of 1,000 kroner per month. While more expensive than on-street parking in the older parts of Copenhagen, the parking in Sydhavn is an attraction for upper-class demographics seeking luxury housing. In some blocks of flats, such as the Metropolis, the apartments are second homes atop an underground garage, and the development also includes docks for sailboats. The next phase, Teglholmen, was also being constructed with 1:100 m^2 parking, with some sections built by 2018 and others anticipated for completion in the 2020s.

By 2007 Copenhagen's redevelopment structure was reordered by merging Ørestad and the harbor redevelopments into a redevelopment authority with a majority (95 percent) of the land ownership allotted to the city of Copenhagen, and the remainder to the national government (Katz and Noring, 2016). Soil excavated during construction of the new tunnel for the Metro Circle Line was taken by truck to the northern part of Copenhagen harbor where it was used to build new land at Nordhavnen (North Harbor). This was then to be sold at a profit to private developers once the land was stable and ready for construction. Nordhavnen, like Ørestad and Sydhavnen, was planned to be built in incremental stages but this time the sequence of development started on land closest to the city center.

With more direct city control, and at the height of the whirlwind of discourses on Copenhagen as an iconic green mobility city, the city acknowledged that too much parking might undermine other green mobility goals, and therefore returned to the Ørestad parking model of the 1990s, albeit with slightly more parking per residential unit. Off-street parking for residences in Nordhavnen includes separated parking structures and no on-street parking at the curb, reflecting some recognition in the city that parking generates more car dependency. The master plan for Nordhavnen calls for less parking but the build-out will take time (Transport Advocate (3), 2016). While more progressive than the parking ratios for Sydhavnen, taking the longer view, the parking policies in Nordhavnen reflect the incremental upward tick in Copenhagen's accommodation of more cars. Moreover, parking fees are also slightly lower at 900 kroner per month (US $138 per month in 2018) (Transport Advocate and Politician (2), 2016). The future build-out of Nordhavnen includes housing for 40,000 persons, and space for 40,000 jobs on land mostly built from landfill in the shallow harbor. If conservatively assuming the average household size is three persons, and the average unit size 80 square meters, then this amounts to almost 5,800 potential future parking spaces in Nordhavnen.

Defenders of parking in developments like Nordhavnen suggest that even with abundant parking, the need for longer-distance car travel might be removed and so there will be less car travel cumulatively within the Copenhagen region (City Finance Executive, 2016). Moreover, there are arguments that car owners in Nordhavnen will use their cars less overall. Yet the planned mode share for Nordhavnen, one-third car, one-third cycling, and one-third public transit means that there will still be a lot of car use relative to the present mode share of Copenhagen. If one-third of trips in and out of Nordhavnen are by car, this pours thousands of cars into adjacent Østerbro and other inner areas, where we have already seen that road space is contested and there is little room for additional car traffic.

In late 2018, as Sydhavnen was approaching build-out and construction of Nordhavnen was underway, the national government's Neoliberal-Right/Conservative mobility coalition, in negotiations with the city's Social Democrats, proposed a massive new development scheme to the northeast of Christianshavn on Amager, in the shallows of the Øresund and across the harbor from Nordhavnen. This scheme, known as Lynettenholmen, was branded in all the routine green aspirations of the previous round of harbor redevelopment, and little information about parking ratios was provided, but tellingly, this scheme included a massive new car-oriented harbor tunnel connecting Amager to Copenhagen's wealthy northern suburbs via Nordhavnen.

We will consider the merits of the tunnel in Chapter 7, but briefly, as currently framed, and because it is bundled with a new mega-project for cars, the Lynettenholmen proposal appears to top off a path-dependent trajectory away from Copenhagen's legacy as an iconic cycling and green mobility capital. While no concrete parking numbers have been suggested, given the isolated nature of this development area, far from any Metro or S-train, and proposed with a massive car-based tunnel, in all likelihood parking will be considerable. As with the previous iterations of redevelopment at Ørestad and the harbor, car parking establishes a path dependency and promulgates assumptions about future residents wanting cars. Unless the political winds in Copenhagen shift, Copenhagen's well-earned legacy of green mobility will be undermined by massive amounts of new car parking.

Revanchist parking and the future politics of mobility in Copenhagen

In the previous section we discussed the implications of off-street parking in the "new" Copenhagen, the areas of the city where most new housing, commercial, and office development is occurring, and emphasize how parking can set a path-dependent trajectory that may undermine Copenhagen's green mobility legacy. Built on formerly industrial, port, and military brownfield land, these new residential and commercial districts are acclaimed as energy-efficient transit- and bicycle-oriented developments and models of sustainable urban regeneration and reurbanization. Yet despite a green veneer, these new developments are all intentionally car-oriented because they are planned and constructed with private off-street parking for new residents and office workers. Contradicting Copenhagen's visions and aspirations as a cycling city and a capital of car restraint, these new developments invite more car owners to the city.

All of this new parking might also invite changes to Copenhagen's politics of mobility, tilting it towards more car-oriented politics resembling Copenhagen's suburbs and Danish national car politics, and obstructing cycling and transit agendas. More to the point, private off-street parking in the new Copenhagen is revanchist—a recovery of lost territory or status—as more neoliberal and conservative thinking about the car accompanies the reurbanization process (see Henderson, 2013, for a more extended discussion of revanchist transit policies). New parking may also be bound up with the demographic transformation of Copenhagen into a wealthier and exclusive city, especially since a key argument in favor of the off-street parking provision is that it is considered necessary for attracting higher-income residents to the city. This might make Copenhagen's Left/Progressive alignment with the "right to the city" political discourse even more difficult to realize (Christiansen and Thomas, 2016).

The new Copenhagen, with tens of thousands of new housing units, new office buildings, public buildings, hotels, Scandinavia's largest conference center, and shopping districts, is a decidedly neoliberal redevelopment scheme. It is foisting the role of an entrepreneurial real estate developer onto the city of Copenhagen in exchange for revenue to finance infrastructure investment and increase tax revenue from new higher-income residents. Presuming that wealthier households will include many families with children, new residential flats are also built larger and more spaciously, with abundant green amenities near by, thus increasing land values further.

While at face value the rationale that higher-income tax-paying households are beneficial to the municipal social welfare system, this new stratum may further essentialize the car in Copenhagen's politics of mobility. For example, many new families in the new Copenhagen might adopt a Liberal Alliance viewpoint that families with children must have cars, and therefore if Copenhagen is to attract new young families, new housing must provide car parking. Shops, schools, and other urban services, it stands to reason, would also need to provide parking. As families shift into car-oriented lifestyles in the new Copenhagen many might also purchase a second car, placing more demands on parking supply and putting more pressure on the municipal government to require more parking as the city adds housing.

Other new Copenhageners might sometimes commute by bicycle or appreciate the livability provided by the spaces of green mobility, but also demand parking for a private car so that they might drive at the weekend to their summer cottage or to chauffer children to soccer matches, etc. (Jensen, quoted in Gössling, 2013). These new families might use their cars sparingly, but the massive amounts of space and building materials needed for storing these cars comes at the expense of other urban spaces, and these households are susceptible to lapsing into more car use, for shopping or when the weather turns wet and cold, for example. These households will also expect parking at the destination of their car trips, affecting future decisions for developments in other parts of Copenhagen.

Sydhavnen's new residents might provide a lens into this car-oriented future scenario. It is relatively easy to drive from Sydhavnen to the regional motorway network, and the connections to the rest of the city for cycling are not particularly good, although new cycle track improvements were introduced in 2018. The infill railway station once proposed for Sydhavnen was never built, and the Metro extension into Sydhavnen is years from realization. It is no surprise then that Sydhavnen, with some of the highest-income precincts in the city, has a much higher car ownership rate than the rest of Copenhagen (Transport Advocate (3), 2016; Transport Advocate and Politician (2), 2016). Many residents are thought to be self-selecting with parking as a determinant for residential location. These new residents are predisposed to want a car and prefer to drive, but might locate outside Copenhagen if parking is not supplied (Transport Advocate (3), 2016). Of particular interest, however, is the emerging concentration of Liberal Alliance voters in Sydhavnen. With parking and luxury housing comes not only demographic change, but political change, and in this case an explicit politics in favor of the car.

If demographic change in the new Copenhagen is accompanied by political shifts, it is important to understand what that looks like. For example, in general, the Right/Conservatives advocate for more off-street parking. Essentializing the car, in local politics Conservatives have opined that people should be able to park a car immediately outside their house, and that new residents will need cars, so parking must be included in new developments (Conservative People's Party, 2013). The Conservative People's Party is joined by the Danish People's Party and the Liberals in insisting that new housing should include at least one parking space per new unit (Transport Advocate and Politician (1), 2016). From a Neo-liberal perspective parking is deemed necessary for economic growth and for attracting wealthier residents to the city (see above). New off-street parking is also necessary because it mitigates the removal of on-street parking as the city grapples with how to balance cycling, transit, and cars. Private off-street parking is also a new market for buying and selling urban real estate—in this case private car storage instead of housing.

Meanwhile the Social Democrats' ambivalence about the car continues with respect to the parking issues. Stakeholders have accused the Social Democrats as having a "double agenda" on parking. On the one hand the Social Democrats boast about Copenhagen's Climate Plan but then maintain that parking must be provided because cars equal economic growth, and the municipality depends on the revenue from car-owning residents to help underwrite social welfare in the city.

Stakeholders familiar with the drafting and promotion of Copenhagen's Climate Plan have even suggested that the plan lacks a parking goal because of a "gag order" imposed on city staff and the discussion of parking (City Climate Planner (1), 2016; City Transport Planners (2 and 3), 2016). Rather than broach the uncomfortable and vitriolic emotional attachment between car drivers and parking spaces, the Social Democrats' city planning discourse only focuses on the "carrots" of better cycling and public transit, shunting the difficult spatial reality of parking and cars off to later generations to confront (City Climate Planner (1), 2016). This is especially troublesome since the first iteration of the Climate Plan in 2012 included congestion pricing, and discussions about parking policy were deliberately sidelined. At the time, it was assumed that continuing to discuss parking would make it impossible to adopt the climate plan (City Climate Planner (1), 2016).

For Copenhagen's Left/Progressives, there is pointed frustration with the Social Democrats, who, some argue, "protect car owners" and think that "all people have cars and want parking." According to the Left/Progressives, the parking politics of the Social Democrats are contradictory to their endorsement of green goals and visions. For example, one urban planner remarked that "The Social Democrats were all-in on carbon neutral discourse but when it came to parking, they invoked a rhetoric that a single mother with two children must have a car and parking," thus stonewalling Left/Progressive efforts to make it safer for both mothers and children to cycle in Copenhagen. Within Copenhagen's Left/Progressive urban planning faction, there is deep frustration that parking in Copenhagen is not being discussed openly or honestly (City Transport Planners (2 and 3), 2016).

Copenhagen's Left/Progressives, and especially the Red-Green Alliance and the Alternative, advocate for less parking, for requiring that whatever parking is built must be placed underground or be invisible, and that the city should be explicit about the cost of parking and the impact that it has on housing prices. Some have also proposed a tax on new private off-street parking (Transport Advocate (2), 2016). The Socialist People's Party opined that Copenhagen needs more parking for bicycles and less for cars, although in 2018 it appeared that the party was willing to back off on the issue of parking for other reasons (Aggesen, 2013). The Left/Progressives point to public opinion surveys of Copenhagen residents and suggest that few Copenhageners complain about a lack of car parking, but many complain about crowding on the cycle tracks. Surveys also show that residents want to remain in Copenhagen because of cycling and livability, not because of car parking (City Transport Planners (2 and 3), 2016). All of this informs the political action taken by the Left/Progressives.

In 2015 Copenhagen's Left/Progressives capitalized on such sentiments and, led by the Red-Green Alliance, spearheaded a reduction of the city's parking requirements when Copenhagen's Municipal Code was revised (City of Copenhagen, 2015c; City Transport Planners (2 and 3), 2016). In this theater of Copenhagen's politics of mobility, the Left/Progressives promoted deeply restrictive parking measures in parts of the city within close proximity to Metro and S-train stations (which cover a substantial part of the city).

Debate centered on off-street parking minimums and maximums for new residences and mixed use developments (housing and shops, etc.) within certain distances of the transit nodes (Transport Advocate and Politician (2), 2016). The Left/Progressives fought for one parking space for every 200 square meters of residential space in areas within 600 meters of a transit station. The eventual political compromise halved this, at 300 meters from stations. If considering that on average the floor area of an apartment in Copenhagen measures 80 square meters, this approximates to one parking space for every two-and-a-half or three housing units.

The Left/Progressives also sought to increase the minimum (and maximum) to one parking space per 300 square meters for housing within 150 meters of a transit station. Again, taking the average apartment size in Copenhagen into consideration, this would amount to approximately one parking space for every four housing units (Transport Advocate and Politician (2), 2016). Very few places in the world have such low parking requirements, with a very small part of San Francisco's inner city having 0.25:1 parking in parts of the Market and Octavia Plan, for example (Foletta and Henderson, 2016; Henderson, 2013).

After compromises were reached, stricter off-street parking standards were adopted with two-thirds of the City Council's approval (Transport Advocate and Politician (2), 2016). New parking minimums were imposed for new housing developments within a short distance (300 meters) of a train station. At stations, the parking minimum was adjusted to 1:250 m^2, or roughly one parking space for every three units of housing (0.36 parking spaces per unit of housing in US parlance), based on new units with a floor area measuring on average 80.5 square meters. This is a progressive minimum relative to any global city comparison, and Copenhagen's Left/Progressives clearly delivered on that score.

In the new redevelopment areas parking was set at 1:200 m^2, which denoted a return to the 1990s Ørestad plan. While on average a typical new flat built in 2014 in Copenhagen had a floor area of about 80.5 square meters, a substantial number of new flats in redevelopment areas such as Sydhavnen approach 100 square meters. Copenhagen's new parking minimums for redevelopment areas equates to roughly one parking space for every two units of housing (if displayed in US parking ratios this equates to 0.4:1 to 0.5:1 parking spaces per housing unit). In citywide mixed-use areas where there is housing and commercial activity, more parking is required for the floor area, at 1:150 m^2 (0.54:1 in US figures).

These tougher off-street parking restrictions reflect the political influence of Copenhagen's Left/Progressives, but they are a compromise, and there are no minimums. This means that, technically speaking, off-street parking must be provided in any new housing development, regardless of location—even atop a new Metro station in the Brokvarterene. Although the ratio of parking to housing units is less than 1:1 (if taking Copenhagen's historic averages of normal floor area), in the public transit- and cycling-rich urban core, parking must be provided, despite the long history of no off-street parking. Copenhagen does cap the amount of parking with maximums of one parking space for each 100 square meters meaning that new houses cannot come with multiple parking spaces for each new unit. But the maximum of $1:100^2$ is substantial, and with much of the new housing constructed in the harbor area and other districts, this effectively means that many new housing units are accompanied by one new parking space.

As previously mentioned, the average size of apartments in Copenhagen has crept steadily upward, and many new luxury units are in the 100-square-meter range. This provision of parking coupled with larger units further suggests an upscaling of the city and potential political shifts in the future. This brings considerable change to the urban core, with more cars, more car-trip generations, and, also, more places where cars enter or exit streets, presenting new conflicts for cyclists and pedestrians. The presence of any additional car traffic in much of Copenhagen is likely to result in more conflict.

However, when comparing Copenhagen to peer cities in Europe or North America, the city's off-street parking requirements are still relatively progressive and credit is due to Copenhagen's Left/Progressive advocates and elected officials (City Transport Planners (2 and 3), 2016). That parking reform efforts have been largely carried forward by the Left/Progressives is not uncommon in peer European cities, but is instructive for Left/Progressives in the United States, where there is a need to acknowledge, recognize, and incorporate parking reform into broader green mobility visions.

Copenhagen's politics of parking is not unique and cities all around the world have very similar politicized debates. In the United States, Los Angeles, New York, San Francisco and even Portland have experienced politically charged backlashes against parking removal and reallocation for the purpose of building cycle tracks, as well as debates over public transit versus car parking. In many European cities, narrow neighborhood streets are crammed with parked cars making it less comfortable to cycle.

In many of Copenhagen's peer green mobility cities such as Amsterdam, Berlin, Hamburg, Munich, Vienna, Zurich, and San Francisco, the Left/Progressives were spearheading parking reform and broader green mobility policies (Foletta and Henderson, 2016; Buehler et al., 2016; Buehler et al., 2017). A common link is that local Left/Progressive politics shaped the parking management policies, and that backlashes against parking management from the right-wing tempered and modified the degree of their implementation. In Vienna, arguably one of the top green mobility capitals in the world, a coalition of left-leaning Social Democrats and Green Party politicians steered the city towards a reduction in car use through parking management and investment in public transit (ibid.). Right-wing conservative elements, while endorsing some forms of parking management as part of the historic preservation of Vienna's signature urban core, have pushed back on the more far-reaching parking plans.

Currently, the Left/Progressives in Amsterdam and Zurich lead the efforts to remove and impose steep increases in the price of parking in those cities (Foletta and Henderson, 2016; Buehler *et al.*, 2016). In Oslo, the Left/Progressives, seeking to reduce car parking and car access in the city center, have promoted car restraint and less parking in the urban core of that city. One of the most widely known car-lite developments in Germany, Vauban in Freiburg, has limited parking because of efforts by Left/Progressives, but the extent of parking reduction was tempered by the more Conservative government of Baden Wurttemberg. Copenhagen's Left/Progressives are not alone in experiencing backlashes from the right-wing on parking.

In the United States, the parking politics of San Francisco stands out as most ground-breaking but are also rife with vitriolic backlash. During the early 2000s San Francisco's Left/Progressive Board of Supervisors (or local city council) adopted some of the nation's strictest off-street parking limits for new housing, and directed revenue from on-street parking towards public transit (Henderson, 2013). Soon afterwards a politically Conservative backlash, couched in a populist "war against cars" narrative, went to the ballot and although unsuccessful, tempered the city's political will to go further on parking reform (ibid., 2015). Politically Conservative car owners (at least in respect to urban planning), many residing in single family homes in the city's outer precincts, as well as merchants, balk at limiting parking in new housing developments. More recently, in late 2018 San Francisco approved the elimination of parking minimums citywide as a combined green mobility and affordable housing strategy. The new zoning was adopted in January 2019 and spearheading the effort was the most left-leaning member of the city council (Sabatini, 2018).

Conclusion: parking policy is cycling policy

Historically, the absence of parking has been fundamental to what makes Copenhagen a green mobility capital. A car is not needed at all to live in Copenhagen, yet for ideological and political reasons, and with a Conservative and Neoliberal tilt, many new residents choose to bring a car. This is likely to be accompanied by a political revanchism that will undermine much of what Copenhagen has come to stand for in terms of green mobility, and could also result in a dampening of the Left/Progressive base as gentrification and displacement in older sections of the city are accompanied by upscaling car-oriented development in the harbor area and Ørestad.

As the next phase of redevelopment of the harbor area and other parts of the city get underway, Copenhagen is at a critical juncture for rethinking off-street parking but also what kind of city it will be in the future. The revision of Copenhagen's Municipal Plan, scheduled to take place in 2019, might provide a window of opportunity to rethink off-street parking in ways that align Copenhagen's cycling and broader green mobility agenda with new development (City of Copenhagen, 2015c). Copenhagen could eliminate minimums, and promote car-free development instead of continuing to accommodate more cars in the "new" Copenhagen. If Copenhagen is to remain a strong cycling city, and revamp its status a global leader, it is time for the city to recognize that parking policy is cycling policy.

References

Aggesen, S. 2013. "Where Are the Cars Parked?" *Politiken*, 21 October.

Andersen, H. T. and Jørgensen, J. 1995. "Copenhagen." *Cities*, 12: 13–22.

Astrup, J. 2018. "Enhedslisten Mayor: Parking Must Be More Expensive." *Politiken*, 8 May.

Astrup, J. 2019. "See Which Bicycle Projects Survive, and Which Get the Knife." *Politiken*, 9 January.

Beatley, T. 2000. *Green Urbanism: Learning from European Cities*. Washington, DC, Island Press.

Buehler, R., Pucher, J., Gerike, R., and Götschi, T. 2016. "Reducing Car Dependence in the Heart of Europe: Lessons from Germany, Austria, and Switzerland." *Transport Reviews*, 37(1): 4–28.

Buehler, R., Pucher, J., and Altshuler, A. 2017. "Vienna's Path to Sustainable Transport." *International Journal of Sustainable Transportation*, 11: 257–271.

Cervero, R. 1998. *The Transit Metropolis: A Global Inquiry*. Washington, DC, Island Press.

Christiansen, F. and Thomas, O. K. 2016. "Der vil ske det samme, som skete på Vesterbro for 15–20 år siden." ("Sydhavn Will Go Through the Same Thing as Vesterbro 15–20 Years Ago"). *Politiken*, 1 August.

Christiansen, F. 2018. "Government to Municipalities Raising Parking Fees: No. "*Politiken*, 31 May.

City of Copenhagen. 2012. *CPH 2025 Climate Plan: A Green, Smart, and Carbon Neutral City*. Copenhagen, Technical and Environmental Administration.

City of Copenhagen. 2015a. *Copenhagen City of Cyclists: The Bicycle Count 2014*. Copenhagen, Copenhagen Technical and Environmental Administration.

City of Copenhagen. 2015b. *2015 First Annual Parking Report*. Copenhagen, Technical and Environmental Administration.

City of Copenhagen. 2015c. *Copenhagen Municipal Plan 2015: Coherent City*. Copenhagen, Finance Administration.

City of Copenhagen. 2015d. *Copenhagen Green Accounts 2014*. Copenhagen, Technical and Environmental Administration.

City of Copenhagen. 2016. *2016 Annual Parking Report*. Copenhagen, Technical and Environmental Administration.

City of Copenhagen. 2017c. *Cycle Track Priority Plan (2017–2025)*. Copenhagen, Technical and Environmental Administration.

City of Copenhagen. 2018a. *Member Proposal on the Introduction of Road Pricing in Copenhagen*. Copenhagen, Copenhagen City Council. Available at www.kk.dk/indhold/borgerrepra esentationens-modemateriale/31052018/edoc-agenda/babf8d9b-aace-414a-999f-818a1a 5951fa/0bfbc6f5-ab91-4bf8-b9e3-34f51d75bc83 (accessed 30 May 2018).

City of Copenhagen. 2018b. *Parking Report 2018*. Copenhagen, Technical and Environment Administration.

City of Copenhagen. 2018c. *Parking Negotiations and Vote, May 7 2018*. Copenhagen, Copenhagen City Council. Available at www.kk.dk/indhold/teknik-og-miljoudva

lgets-modemateriale/07052018/edoc-agenda/ea6edf64-ad00-428e-bb6b-7e24dee1eacf/ 07ec48fd-f127-4da8-b60c-196326bf7097 (accessed 15 June 2018).

City of Copenhagen and State of Green. 2014. *CPH 2025 Climate Plan: Solutions for Sustainable Cities*. Copenhagen, City of Copenhagen and State of Green.

Conservative People's Party of Copenhagen. 2013. "There Must Also Be Room for Cars." *Politiken*, 29 January.

Conservative People's Party of Copenhagen. 2017. *Transport Policies*. Copenhagen, Conservative People's Party. Available at https://konservative.dk/politik/politikomraader/tra nsportpolitik/?cn-reloaded=1 (accessed 28 July 2018).

Danish Architecture Center. 2015. *Guide to New Architecture in Copenhagen*. Copenhagen, Danish Architecture Center.

Enhedslisten. 2018. *A Budget Plan with Black Fingerprints*. Copenhagen, Enhedslisten.

Foletta, N., and Henderson, J. 2016. *Low Car(Bon) Communities: Inspiring Car-Free and Car-Lite Urban Futures*. New York, Routledge.

Gehl, J., and Gemzoe, L. 2003. *New City Spaces*. Copenhagen, Danish Architectural Press.

Gössling, S. 2013. "Urban Transport Transitions: Copenhagen, City of Cyclists." *Journal of Transport Geography*, 33: 196–206.

Hansen, A. L., Andersen, H. T., and Clark, E. 2001. "Creative Copenhagen: Globalization, Urban Governance and Social Change." *European Planning Studies*, 9: 851–869.

Henderson, J. 2013. *Street Fight: The Politics of Mobility in San Francisco*. Amherst, University of Massachusetts Press.

Henderson, J. 2015. "From Climate Fight to Street Fight: The Politics of Mobility and the Right to the City." In Julie Cidell and David Prytherch, eds. *Transport, Mobility, and the Production of Urban Space*. Oxford, Routledge.

Katz, B. and Noring, L. 2016. *The Copenhagen City and Port Development Corporation: A Model for Regenerating Cities*. Washington, DC, Brookings Institution.

Kodukula, S., Frederic, R., Jansen, U., and Amon, E. 2018. *Living. Moving. Breathing: Ranking of European Cities in Sustainable Transport*. Wuppertal, Wuppertal Institute.

London School of Economics and Political Science (LSE). 2014. *Copenhagen: Green Economy Leader*. London, LSE.

Newman, P., Beatley, T., and Boyer, H. 2009. *Resilient Cities: Responding to Peak Oil and Climate Change*. Washington, DC, Island Press.

Newman, P., and Kenworthy, J. 2015. *The End of Automobile Dependence: How Cities are Moving Beyond Car Based Planning*. Washington, DC, Island Press.

O'Sullivan, F. 2016. *Even Copenhagen Makes Mistakes*. Philadelphia: Next City. Available at https://nextcity.org/features/view/copenhagen-affordable-housing-sustainable-cities-m odel (accessed 25 December 2018).

Pedersen, P. A. 2018. "The Unity List: It Makes No Sense to Cut so Markedly Down on Social Housing Projects in Copenhagen." *Politiken*, 9 September.

Sabatini, J. 2018. "Minimum Parking Requirements on Their Way Out in SF." *San Francisco Examiner*, 4 December.

Sadik-Khan, J. and Solomonow, S. 2016. *Street Fight: Handbook for an Urban Revolution*, New York, Viking.

Saehl, M. 2013. "Car Traffic on Nørrebrogade Has Fallen by 60 Percent." *Politiken*, 21 February.

Shoup, D. 2005. *The High Cost of Free Parking*. Chicago, Planners Press, The American Planning Association.

State of Green. 2016. *Sustainable Urban Transportation*. Copenhagen, State of Green.

Interviews held by Jason Henderson from 2015 to 2017 with transportation experts in Copenhagen

City Climate Planner (1), 2016.
City Finance Executive, 2016.
City Transport Planner (1), 2016.
City Transport Planners (2 and 3), 2016.
City Transport Planner (4), 2016.
City Transport Planner (5), 2016.
City Transport Exec (1), 2016.
Transport Advocate (2), 2016.
Transport Advocate (3), 2016.
Transport Advocate and Politician (1), 2016.
Transport Advocate and Politician (2), 2016.
Transport Consultant, 2015.
Transport Consultant (2), 2016.
Transport Consultant (3), 2016.
Transport Scholar (1), 2016.

7

FROM THE HARBOR TUNNEL TO THE METRO CITY RING

What kind of city?

What kind of city?

By now it should be obvious that there is a contradiction in Copenhagen's status as a green mobility capital. Copenhagen's enthusiastic admirers from almost all political stripes celebrate Copenhagen as the iconic cycling city. Much of that enthusiasm is warranted since within Copenhagen most trips to places of work and education—62 percent—are by bicycle, and for all trips the rate is 29 percent. Driving accounts for just 9 percent of all trips made within the city. This is impressive by any global comparison. Public transit is also on solid footing relative to many peer cities around the world, and the city is very walkable. Yet "Something is rotten in the state of Denmark." Car ownership and car driving are increasing, especially in Copenhagen's suburbs, and mostly for trips that either start or end in the city. At the same time, cycling is plateauing.

Copenhagen's ambitious 2025 Climate Plan reflects this contradiction. When it was first drafted in 2012, the toll ring was supposed to help to reduce the proportion of driving trips to from 34 percent to 25 percent. Plans for the toll ring were scuttled. Next, city planners hoped that parking management, as a form of car restraint, could help to achieve this target. As we saw in the previous chapter, the politics of parking is frustratingly difficult. Copenhagen then moved the goalposts, and in 2016 the 25 percent target for car trips was deferred until 2025. Even with the ten-year extension the only real plan for reducing car emissions by 2025 was to alleviate increased car emissions with questionable tactics by burning biomass and then offsetting that with investments in utility-scale wind energy. Expanded car restraint was omitted from the climate plan.

In 2018 several other important and interconnected flashpoints swirled around Copenhagen's politics of mobility and some of these might make the climate goals for Copenhagen even more elusive. Copenhagen's biggest street fight in generations, the proposed harbor tunnel and Eastern Ring Road, was aggressively promoted by the Neoliberal-Right/Conservative coalition. This mega-highway project, which has been debated for decades, includes a four-lane, 12-kilometer (7.5 mile)-long underground motorway around eastern sections of Copenhagen (see Figure 7.1). If built, this ring would complete the circle for an orbital motorway around Copenhagen.

Alongside the proposed tunnel and bypass there is an associated debate about making parts of the center of Copenhagen car free, which fits in with Copenhagen's green mobility and climate goals. The Left/Progressives have for decades been advocating the establishment of a car-free center, but new proposals have made the car-free center conditional upon construction of the harbor tunnel and Eastern Ring Road in an attempt to creatively trade new car spaces created by the tunnel and bypass with less car space in the city center (Kirkegaard, 2018).

Meanwhile, there are important debates about the future of public transportation and how public transportation interacts with both cycling and the car. The venerable regional S-train network, which has proved very successful, suffers from capacity and congestion problems, the public bus system is frequently stuck in traffic, and the new Metro Ring City line, scheduled to open in 2019, may not have been well coordinated with Copenhagen's bicycle and car policies. Once opened, the new Metro City Ring could siphon off cyclists while making it easier to drive in the urban core. Public transit is a keystone for a green mobility city and, when harmonized with cycling, offers a very different vision of the city's future compared to the harbor tunnel and Eastern Ring Road. Crosscutting all of this are recent proposals to remove the elevated Bispeengbuen motorway, but with caveats that preserve the same car capacity and lock in future on-street parking revenue. Copenhagen is clearly at a unique crossroads regarding what kind of city it may become. Our work would not be complete without touching on all of the moving pieces.

The harbor tunnel debate

The proposed four-lane, 12-kilometer (7.5 mile)-long harbor tunnel, sometimes interchangeably described as the Eastern Ring Road, would complete an orbital motorway around Copenhagen. As shown in Figure 7.1, from the north the tunnel would connect to the Northern Harbor Road Link (Nordhavnsvej), which is effectively a short four-lane motorway extending from the Helsingor motorway to just west of Nordhavnen. The harbor tunnel would drop under Nordhavnen and then swing southeast across the entrance to Copenhagen harbor, before connecting to a derelict industrial district called Refshaleøen.

With the harbor tunnel as a centerpiece and catalyst, the former shipyards and adjacent shallow waters around Refshaleøen have been targeted for large-scale landfill and conversion into a massive housing and employment center known as Lynetteholmen (the name refers to a nearby marina). As with the harbor redevelopments at Nordhavn and Sydhavn, landfill for this development would come from the harbor tunnel excavation (and possibly from another Metro line

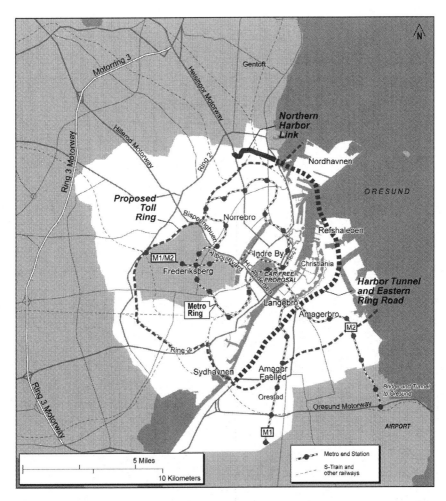

FIGURE 7.1 Map showing the harbor tunnel proposal, the new Metro City Ring and the proposed toll ring described in Chapter 5.

Source: Map by Michael Webster.

extension). Following the route mapped in Figure 7.1, south of Refshaleøen the proposed harbor tunnel alignment bypasses east of Christiania and bends west-southwest underneath the dense inner city portion of Amager, paralleling the Inner Harbor channel and then finally surfacing at a junction with the Øresund motorway. This completes the motorway ring around Copenhagen, since continuing westward connects to Ring 3, Copenhagen's western motorway bypass. To the east, the Øresund motorway connects with Copenhagen international airport and to Sweden via the Øresund bridge and tunnel.

Proponents of the harbor tunnel/Eastern Ring Road, including the Social Democrats, the Liberals, and the Liberal Alliance in the Neoliberal faction, and the Conservative People's Party on the Right/Conservative side, argue that the tunnel and bypass are needed to ensure the economic growth of both Copenhagen and Denmark, and in order to provide better connections to Sweden and to Copenhagen's international airport at Kastrup. The Danish People's Party, it should be noted, is not fully committed to this plan, but nor is it openly opposed to it. This nuanced stance reflects the view that economic development should be directed away from Copenhagen. It is not an anti-road position (Danish People's Party, 2018).

The Left/Progressives in Copenhagen are opposed to the harbor tunnel and Eastern Ring Road. The Red-Green Alliance has led objections to the mega-project for the past decade, arguing that the roadway would invite more cars into the city and ultimately undermine cycling, public transit, and Copenhagen's broader climate goals. During 2017 and 2018 the Alliance was accompanied by the Socialist People's Party, the Alternative, and the Social Liberals, a predictable alignment, in voicing concerns about the tunnel and bypass scheme.

Some Left/Progressives view the harbor tunnel and Eastern Ring Road as deeply inequitable because the scheme commits future financial and material resources to a high-speed roadway used primarily by a car-driving elite traveling from the wealthy "whiskey belt" (Copenhagen's northern suburbs) to the airport or to Sweden (Colville-Andersen, 2012; Transport Consultant, 2016). Furthermore, some Left/Progressives, especially members of the Red-Green Alliance and the Alternative, have joined together in protesting the impact that the tunnel and adjacent development would have on Amager Faelled, a green space located at the point at which the tunnel's southern portal would surface (see Figure 7.1). If built, the tunnel locks Copenhagen into a more car-oriented development pattern, at least on Amager and along the alignment of the roadway, and especially when considering the provision of off-street parking in the adjacent redevelopment areas.

A brief history of the harbor tunnel and Eastern Ring Road

The harbor tunnel and Eastern Ring Road are the latest iterations of a long political struggle to build an eastern bypass around Copenhagen (Homann-Jesperson *et al.*, 2001; Transport Scholar (1), 2016). The tunnel was mapped in Copenhagen's regional motorway plan of the 1960s as part of the Ring 3 motorway. Owing to the needs of maritime shipping it was always envisioned as a tunnel rather than a bridge (Homann-Jespersen *et al.*, 2001). The aim of roadway planners was to have a complete limited access beltway around Copenhagen, in much the way that cities in the United States such as Atlanta, Denver, or Washington, DC, have beltways, as do Berlin, London, and Paris in Europe. With so many other world cities completing beltways, and Copenhagen's intention to have one of its own, why was the tunnel/bypass not built?

As previously discussed in Chapter 2, Left/Progressive political opposition to cars and new urban environmental awareness in Copenhagen resulted in decisions to scale back Copenhagen's motorway plans (Pineda and Vogel, 2014). This included cancellation of plans to construct motorways through Copenhagen along the lakes, but an eastern bypass motorway remained on the planning maps for Copenhagen. Traffic planners insisted that the bypass would be needed to move traffic around Copenhagen. Following the oil crisis and the deindustrialization of Copenhagen in the 1970s, planning for the eastern bypass stalled for almost two decades (Danish Town Planning Institute, 2004).

In 1993 debate over the harbor tunnel and the Eastern Ring Road started up again (Homann-Jesperen *et al.*, 2001). The shipbuilding firm Burmeister & Wain (B & W), with a major shipyard on Refshaleøen (which is poised for redevelopment in the current harbor tunnel scheme), argued that the tunnel was necessary to keep the shipyard functioning and facilitate shipyard expansion. Copenhagen's working-class political history with the shipyard played into the tunnel proposal, and the Social Democrats considered making the tunnel a priority. The Ministry of Transport set up a study committee in 1994 to evaluate tunneling alignments.

In 1995 the *East Harbor Study Report* was published and urged the construction of a harbor tunnel because, it was argued, a harbor tunnel would help to preserve industry and jobs as well as open up other areas to redevelopment. It would also remove car traffic from the city center. The 1995 study examined traffic impacts and speculated that the road would attract many new car trips, not just the existing demand, but it also suggested that the bypass would reduce traffic in the city center by 35 percent. Yet while reducing cars in the urban core was a popular idea for the Left/Progressives, the tunnel scheme would simply shift the traffic around, and would not result in reduced car use or car ownership. New car trips attracted to the tunnel would enter and exit the motorway in Østerbro to the north and Amager to the south, but while this would remove some cars from the center, these two areas would experience increased traffic. These same issues have arisen in more recent iterations of the harbor tunnel and Eastern Ring Road.

In 1996 B & W declared bankruptcy and the shipyards were shut down, thereby changing the nature of the tunnel and bypass scheme. Now creditors, including investors in the Refshaleøen Property Company, holding B & W debt advocated for the harbor tunnel as part of a redevelopment scheme centered on reurbanized housing and commerce. The Social Democratic Party grappled with arguments that the road was necessary for the economic development of Copenhagen. Economic growth such as new upscale housing would bring much needed tax revenue to the city, and underwrite social welfare. In 1996 the city and the Ministry of Transport agreed that they would jointly build the tunnel by using tolls to pay off the construction costs and provide revenue for operating the tunnel.

By 2000 this revised version of the harbor tunnel was defeated at the City Hall. The Red-Green Alliance, whose popularity as a major party in the city was increasing, joined the Social Liberals and the Socialist People's Party in opposition, and residents of upper-class Østerbro joined forces with working-class Amagerbro to convince the City Hall (and the Social Democrats) to table the tunnel proposal. Opponents argued that the tunnel would create more traffic problems than it solved, and that it would undermine the concurrent investment in the new Metro line to Ørestad (which was under construction). The right-wing Danish People's

Party, which was not very prominent in Copenhagen, but was expanding nationally, also opposed the tunnel based on the premise that economic investment should be steered away from Copenhagen (Homann-Jespersen *et al.*, 2001).

This round of politics set back, but did not defeat the harbor tunnel and Eastern Ring Road. Rather, it heralded the beginning of two more decades of debate. In the early 2000s consultants proposed that prefabricated tubes should be dropped beneath Copenhagen's harbor and that the harbor floor should be used as a right of way. Pillars and structures supporting existing bridges, the new Metro tunnels running under the harbor, and other pre-existing infrastructure made this idea technically difficult and extremely expensive to implement. It would also have been necessary to dredge up the layers of toxic soil containing mercury buried beneath the floor of the harbor, which would compromise the recent clean-up operation (Transport Scholar (1), 2016). Other shorter tunnels with different alignments and crossing points were vetted but never got far.

However, one section of the overall scheme was built. The Northern Harbor Link Road, highlighted in Figure 7.1, went ahead with less controversy than the harbor tunnel. Approved by the city of Copenhagen in 2010, and built between 2011 and 2017, the Northern Harbor Link was the largest road infrastructure project to be built in Copenhagen for the past 50 years, and signaled the end of the era of Copenhagen's reluctance to build new highways. The new road was designed to provide traffic relief for the streets in nearby Østerbro and Gentofte, an affluent suburb to the north. Completed in 2017, the road was even endorsed by some Left/Progressives because it provided a way of managing the trucks heading to the port and Nordhavn. Many of these trucks drove through streets in Østerbro on their way to dump soil as part of the building land for the Nordhavn development (Transport Scholar (1), 2016). Built to follow the route originally designed for rail tracks, the 2 billion kroner (US $325 million in 2016) roadway included a four-lane, 3-kilometer bypass from the Helsingor motorway to the entrance of Nordhavn at Ring 2. Looking at the map (see Figure 7.1) it is obvious that the Northern Harbor Link is effectively the beginning of the harbor tunnel and Eastern Ring Road scheme.

While the Northern Harbor Link was being planned and built, the next round of the harbor tunnel and Eastern Ring Road debate was reignited when, in 2009, the Danish parliament directed the Ministry of Transport to include the project as part of a broader examination of ways of bypassing Copenhagen by road and rail (Danish Ministry of Transport, 2013). The aim of these bypass studies included ensuring that Copenhagen was a competitive pole for North European growth, but also to provide more immediate and localized relief of car and truck traffic in the urban core of Copenhagen, and in addition to bolster the economic viability of the Amager urbanization schemes (such as that at Ørestad) while taking pressure off of the existing harbor bridges, all of which were experiencing traffic congestion by the mid-2000s.

The Strategic Analysis Report on the harbor tunnel screened eighteen possible tunnel and bypass alignments and narrowed the options down to the single "preferred" option, which is shown in Figure 7.1 (Danish Ministry of Transport, 2013). The 12-kilometer (7.5 mile) preferred alignment was scrutinized further in a feasibility study that was initiated in 2017 and which was expected to be completed in 2019. This feasibility study was jointly commissioned by the Ministry of Transport (headed by the Liberal Alliance) and the Copenhagen Municipal Government (headed by the Social Democrats). Substantial funding for the feasibility study was also provided by the company that owned the former shipyards on Refshaleøen.

If the feasibility study is convincing, and Copenhagen's City Council votes favorably, the next phase of the project requires a detailed and lengthy environmental report, which might take several more years. If Copenhagen's politics of mobility is still favorable towards the tunnel after this environmental study, construction of the harbor tunnel and Eastern Ring Road could commence by 2024 at the earliest. The harbor tunnel and Eastern Ring Road would be completed in 2035 and soil excavated from the tunnel would be directed at the Lynetteholmen development scheme.

During Copenhagen's municipal elections in 2017 the harbor tunnel and Eastern Ring Road were debated, together with an array of other mobility issues such as cycling, electric cars, and congestion pricing (Risbol, 2017). The Liberal Alliance and the Liberals pressed for the toll tunnel. At a mayoral candidates' forum in November 2017, the Liberals argued that the harbor tunnel provided an optimal solution for congestion and that the government had no right to force people out of their cars (assuming that without the tunnel car restraint policies would be tightened) (ibid.). The Liberal party's candidate for the leadership also suggested tunneling another bypass motorway into Amager from the Bispeengbuen and underneath H.C. Andersens Boulevard (described later in the chapter).

The Social Democrats claimed that the tunnel would result in cleaner air, safer roads, and would remove car and truck traffic from the city's streets. The party also campaigned that the new tunnel and bypass would make it possible to create space in the city center for more cycling and walking, suggesting a new car-free or car-lite proposal might be a condition for the construction of the tunnel and bypass to go ahead.

The Red-Green Alliance and the Socialist Peoples' Party opposed the tunnel. The Left/Progressives suggested that instead of building the toll tunnel Copenhagen should build a light rail system and widen cycle tracks. As with previous rounds of debate, the Danish People's Party, which has its core constituency beyond the municipality of Copenhagen, expressed skepticism that this might fail to "balance Denmark." This is consistent with the notion that more commerce and development should be directed away from Copenhagen.

By 2018 the harbor tunnel and Eastern Ring Road were a catalyst for additional bonding between Social Democrats and the Neoliberal faction at the City Hall, and the broader Neoliberal-Right/Conservative mobility coalition became even more sharply in favor of cars. In late 2018 a group of national political leaders from the Liberals (the Danish Prime Minister), the Liberal Alliance (Minister of Transport) and the Conservative People's Party (the Minister of Finance) joined with the Social Democratic Lord Mayor of Copenhagen to announce a new political push for the harbor tunnel and Eastern Ring Road (Hagemeister and Jensen, 2018).

The new scheme called for more land to be filled at Nordhavn plus 2 square kilometers of landfill just off Refshaleøen for what would be called Lynetteholmen. The new land would be protected from sea level rise by a new storm surge defense system. The tunnel and bypass roadway would be debt-financed, with payment from toll collections and from the sale of newly built land. The Liberal Alliance welcomed the arrangement of tolls and land sales as a way of building the road "for free" with no need for government subsidies. In order to defuse political opposition anticipated from the left, the project was rebranded to emphasize affordable housing. Lynetteholmen's potential 35,000 housing units, coupled with a 25 percent inclusionary housing quotient, was marketed as a solution to Copenhagen's housing crisis, even though it would not be completed until 2070 (Danish Ministry of Transport, 2018; Kraul, 2018; Norup, 2018; Hagemeister and Jensen, 2018).

It remains to be seen whether the tactic of rebranding the tunnel and bypass as a housing program has worked. In an important City Hall resolution to affirm that the city would continue to participate in the Lynetteholmen scheme, the Social Liberals and the Socialist People's Party peeled away from the Red-Green Alliance and the Alternative, the only two remaining parties to unequivocally oppose the harbor tunnel and Eastern Ring Road (Kabell, 2018). This led to accusations from the Red-Green Alliance that the other left-wing parties had forgotten their resistance.

Car restraint and the harbor tunnel

The insertion of housing into the harbor tunnel and Eastern Ring Road proposal muddies the right to the city discourse promulgated by Copenhagen's Left/Progressives, and might prove to be an effective tactic for diffusing opposition to the tunnel and bypass. Also blurring the lines of green mobility, Copenhagen's Social Democratic Lord Mayor seeks to make parts of Copenhagen's center car free or car lite on condition of building the tunnel and bypass (Kirkegaard, 2018). Here the tunnel proponents invoke a fashionable urban planning theme supporting urban roadway tunnels as "an efficient way to create more elbow room for citizens," because tunnels free up road space on the surface (Skov, 2018). A similar argument is made to remove the elevated Bispeengbuen and replace it with a tunnel.

There are three factors to consider when examining the connection between the harbor tunnel and making the center of Copenhagen car free. First, is the Social Democrats' car-free proposal genuine or is it simply a bargaining chip? Some Left/Progressives have accused the Social Democrats of green (mobility) washing (Transport Advocate (3), 2016). The Red-Green Alliance and other Left/Progressives have long supported a car-free center, which the Neoliberal-Right/Conservative coalition persistently rejected. The Social Democrats' 2018 harbor tunnel proposal, which outlined the conditions necessary for the creation of a car-free center, was therefore vague and suspect.

This distrust of the genuineness of the 2018 proposal was particularly salient because in the previous year the Social Democrats had helped to block a pilot car-free proposal (Astrup, 2017; *Politiken*, 2017; Mork, 2017). This proposal was the result of a consultation between the Technical and Environmental Committee (headed by the Red-Green Alliance), the police, and the military intelligence services, who were concerned about the possibility of a terrorist truck attack being perpetrated in Copenhagen following a series of violent truck attacks in Germany, France, and Spain in 2016. The Red-Green Alliance, which had long promoted a car-free city center, was accused of abusing power and fearmongering by conflating the car-free proposal with public concern over terrorism (Astrup, 2017). The Left/Progressives retorted that the idea was also promoted by some members of the police and terrorist security specialists.

A second consideration regarding the interplay between the harbor tunnel and a car-free center is buy-in from Right/Conservatives and even some members of the Liberal party. While the Social Democrats have promised to make the city center car free or at least car lite, the party's right-wing allies have balked at the proposal. The Right/Conservatives objected to the 2017 car-free proposal described above based on arguments that it would damage retailers and tourism. In a pointed response to the Social Democrats' car-free proposal in 2018, one Liberal politician described banning cars as "Stalinist" (Kirkegaard, 2018). The Social Democrats, who hold power in Copenhagen, only do so because of the support of small but vocal pro-car parties like the Liberals, the Liberal Alliance, and the Conservative People's Party. This leads some to consider that the promise of a car-free center is an empty one, and that whatever means are implemented will not truly reduce car access to the core.

A third factor, the relationship between toll rates and the car-free center, complicates the Neoliberal arguments of the pro-tunnel camp. The tunnel will need to collect tolls in order to be financially viable. In 2018 a 20 kroner (about US $3 in 2018) toll was proposed for private cars and an 80 kroner (about $12 in 2018) toll for trucks. This rate was based on analysis conducted by the Ministry of Transport in 2013 which suggested that a higher toll would discourage use of the tunnel, and therefore the toll rate was set lower in order to attract the highest amount of cars and trucks.

Advocates of the tunnel already know that the 20 kroner toll rate is insufficient to pay for the construction of the road. To balance the shortfall in toll revenue, financing of the tunnel was connected to the sale of harbor land built from soil removed during excavation of the tunnel, similar to the Metro development schemes. Yet that is also not enough to ensure that the tunnel is financially viable. There must be a guarantee of captive motorists and truckers who pay the toll (Kirkegaard, 2018).

In this light, car restraint in the city center is not a gift to the city. It is required to make the harbor tunnel and Eastern Ring Road financially viable. Yet, as suggested above, the Right/Conservatives will oppose this with a populist thrust, arguing that it would not be fair to force drivers to pay a toll when they could cross town for free via other modes of transport. Given past politics, the toll rates and car-free center will undoubtedly be a fertile Right/Conservative flashpoint.

Shutting off the Ring 2 or other city center cross-town arterials in order to pay for a new road would probably ignite more street fights, although it would also probably force many cars into the tunnel.

Introducing lower tolls merely to placate motorists could, however, further undermine the self-financing scheme for the tunnel and bypass. This means more of the harbor land must be sold at a higher profit, and this could dampen the potential for more affordable social housing or other community purposes. Furthermore, if the harbor land becomes even more necessary for financing the road, and the revenue of the sales restricted only to the road, other worthy public works, such as improving the S-train or regional bus system, cannot be funded out of the scheme. Additionally, no concrete figures have yet been presented relating to the cost of protecting all of these developments from sea level rise. Copenhagen could end up with a very expensive road on land that loses its value due to being a risky investment.

Beyond the complicated dance between proposals for a car-free center, toll rates, and potential populist political backlash, there is the matter of interchanges along the proposed harbor tunnel and Eastern Ring Road alignment. In this case the bypass function of the road conflicts with the temptation to attract more toll revenue by enabling more access points, potentially turning the scheme into a local roadway instead of a proper bypass. This leads some observers to ask, if the tunnel proponents insist that the region needs a bypass, would not turning it into a local road defeat that purpose? Once again, the toll rate complicates this.

Traffic models conducted by consultants in 2013 suggest that if the tunnel excluded interchanges and was only accessible at the northern and southern portals, toll revenues and the cost-benefit ratio of the tunnel would render the tunnel unnecessary from a traffic perspective. This means that the models predicting upwards of 56,000 cars using the tunnel in the future are assuming that much of the traffic would be locally generated for trips from the suburbs into the city or from the city to the suburbs. Hence, for the project to be feasible from an economic perspective, some kind of access along the line of the tunnel would be necessary.

The provision of multiple interchanges, especially on Amager, close to the city center and to Ørestad, contradicts the promises that the harbor tunnel and Eastern Ring Road will remove cars from Copenhagen's surface streets (Transport Scholar (1), 2016). Cars will need to use the nearby surface streets in order to access the motorway tunnel. Everything points to the reality that the harbor tunnel and Eastern Ring Road will act as a magnet for more cars, thus overwhelming the surrounding streets in Amager. Even if some sort of limited car-free center were politically acceptable, the scheme is, at the core, a vision of a car-oriented city.

The cycling and public transit vision

Public transit, if harmonized with the bicycle, offers an alternate vision to the car-oriented harbor tunnel and Eastern Ring Road. As described in Chapter 1, public transit is a keystone for making it possible to sustain a large cycling population in Copenhagen. Copenhagen's S-train and the bicycle fit hand-to-glove and this symbiosis offers a good template for other cities to follow with regards to cycling and transit. Copenhagen's S-train demonstrates that if public transportation is designed to be supportive of cycling, and if cycling is made to be supportive of public transportation, even higher rates of cycling and public transit ridership can occur (Pucher and Buehler, 2012; Nielsen et al., 2018). This symbiosis also enables Copenhagen to have a high rate of car free households (70 percent) and in Copenhagen even families with children can live comfortably and affordably without a car.

There is great potential for public transit to be more harmonious with cycling and thus expanding the armature for a car-free and car-lite vision of Copenhagen. Roughly 89 percent of residents within the municipality of Copenhagen, and 64 percent of suburban residents have good access (EU Study of Transport Agencies in Movia, 2016). Yet to shift car trips in Copenhagen and in the region generally, public transit must be "irresistible," to borrow from the title of an important paper on urban transit (Pucher and Kurth, 1995). It must be easily accessible, comfortable, dignified, convenient, and reliable.

For cycling in Copenhagen, the S-train comes closest to irresistible but also faces challenges, especially with overcrowding and capacity issues due to its success. Copenhagen's other public transit systems, Movia (the bus system), the newly expanding Metro, and a smattering of other rail services offered by the Danske Statsbaner (DSB—Danish National Railway), have many challenges, both for cyclists and for non-cyclist transit users, that must be addressed if the region's public transportation can help to provide an alternative vision to that presented by the supporters of the harbor tunnel and Eastern Ring Road.

Harmonizing cycling and regional rail

Copenhagen's S-train system is an attractive and stable regional electric rail system that compares favorably to many peer cities around the world (Danish Transport and Construction Agency, 2016; see also Cervero, 1998). Beginning in 1934, the S-train was separated as its own operating unit from the DSB (but still owned by the government) and electrified as a commuter rail workhorse for the region. Two lines were electrified in the 1930s and this became the outline for the 1947 Finger Plan, with Copenhagen as the palm of the hand. In 2018 the total length of Copenhagen's S-train network was 169 kilometers (105 miles) and there were eighty-five stations. The trains operate in a radial pattern about 40 kilometers (25 miles) out from Copenhagen Central Station and carried 116 million passengers in 2016 or between 315,000 and 320,000 passengers a day (Statistics Denmark, 2018).

Along with the S-train there is a separate Øresund railway, an electric railway that hugs the coast on both the Danish and Swedish sides of the Øresund Channel, connected via the Øresund bridge and tunnel. It serves the cities of Copenhagen (and Copenhagen international airport) as well as Helsingor to the north of

Copenhagen, and Lund and Malmö in Sweden. In addition, the DSB operates intercity passenger trains that serve important suburban employment centers such as Høje Taastrup and major towns on Zealand that are beyond the S-train's geographic range.

These are separate operating units from the S-trains, with their own political backstories. The Øresund railway, for example, is jointly owned by DSB and the Swedish railways, and has different operating characteristics (National Transport Planner, 2017). Yet despite being different systems, all of these railways have the same potential to harmonize with cycling and provide an armature for compact station-oriented city planning.

During the past two decades ridership on S-trains has gradually increased and this has been due to improvements in speed and quality of service, as well as allowing free bicycle carriage on board all trains (Danish Transport and Construction Agency, 2016). Copenhagen's S-train system is notably family friendly as well. The success of the S-train is, as with Copenhagen's congested cycle tracks, attracting even more cyclists to the trains, leading to overcrowding and missed trains when a cyclist cannot board. Higher frequencies and longer trains may help, but in the short term, self-sorting with some cyclists leaving their bicycles at the station will be necessary. This means that station parking is also important, and as described in Chapter 3, some stations have parking problems. These two flash-points, however, are seemingly minor and easily fixable. When considering what kind of city Copenhagen might be in the future, there are some bigger problems for the trains.

Copenhagen's S-train and cycling symbiosis is more profoundly undercut by the weakening of the transit-oriented station proximity law that was much celebrated by Cervero (1998) and the London School of Economics and Political Science (2014), and was a core tenant of the Finger Plan. During the late twentieth century Denmark granted land use planning controls to regional administrations (rather than municipalities) and the station proximity law was a foundation of the Finger Plan. This law required that all new moderate-to-large office and commercial developments in the Copenhagen region must be located within 600 meters of a rail station (Andersen and Jørgensen, 1995; Cervero, 1998; Galland and Enemark, 2012).

The station proximity law was "gutted," as one knowledgeable planning expert put it, between 2001 and 2011. The Neoliberals and the Right/Conservatives, with the Liberals at the helm and the Danish People's Party and the Conservative People's Party assuring a majority coalition during their tenure, weakened these laws (Galland and Enemark, 2012; Transport Scholar (1), 2016). The changes stripped land use planning from the regional authorities and devolved power to the municipalities. The municipalities had more discretion to waive the station proximity law (now only advisory) at the request of a corporation or real estate developer (Andersen, 2008; Galland and Enemark, 2012).

The decidedly neoliberal rationale to gut the station proximity law was that the market should determine location decisions, not the government. A secondary and related dynamic was to enable private developers and firms the choice to provide abundant parking and locate where they deemed fit for their own economic preferences, which could mean adjacent to a motorway and far from an S-train station.

Making the station proximity law toothless threatens to undermine the long-term potential of the S-train-cycling symbiosis. As more commercial and office developments sprawl away from stations and away from the Fingers in metropolitan Copenhagen, this induces more car travel and less commuting by bicycle and train. Moreover, as more cars proliferate in Copenhagen's suburbs, cycling itself becomes less and less comfortable and attractive.

This in turn will create more pressure by suburban motorists for car access into Copenhagen, and for new car-oriented Copenhageners to drive out of Copenhagen to dispersed employment centers. There is no doubt that the S-trains will remain full and critical for the region. Yet the gutting of the station proximity law, the continued promotion of the car by the Neoliberals and the Right/Conservatives, and the potential construction of a massive new harbor tunnel and Eastern Ring Road combined will detract from green mobility visions. Copenhagen and the region might instead invest in a better and more expansive regional rail system that is harmonized with cycling, coupled with re-establishing stronger land use controls.

Movia, Copenhagen's neglected bus system

Movia, which means "moving," is Copenhagen's regional public bus system and has the highest share of transit trips in the city (Movia also operates the ferries in the harbor). In 2017 approximately 250,000 daily bus trips started or ended in Copenhagen and Frederiksberg, out of roughly 615,000 trips throughout Zealand (Movia, 2018). Regionally Movia had 1,450 buses in operation across all of the Island of Zealand in 2016, which roughly equates to the expanded Copenhagen metropolitan area of 2.6 million people (ibid., 2016). Of all the disparate public transit systems serving Copenhagen, Movia has the most expansive geography.

Movia is governed jointly by forty-five municipalities and Zealand's two administrative districts—Capital Region and Zealand (Movia, 2016). As with broader trends of liberalization and privatization throughout Denmark and Europe, Movia has a semi-neoliberal streak. For example, Movia outsources work. Separate companies hire drivers and bid for the right to operate on specific bus lines established by Movia. These companies own their buses but must standardize them to the regional fleet, and follow set routes and timetables established by Movia's governing board.

Compared to the Metro and the S-train, the buses' share of trips have been stagnant or declining. Between 2013 and 2018 bus ridership slipped by just 0.3 percent annually in the region, with a much higher decline within Copenhagen (Movia, 2018). Since Copenhagen underwent population and economic growth, this implies that some new car trips were shifted from the bus. Slower transit and cutbacks on service leads to more driving.

To stave off declines in bus ridership, Movia proposed a transit priority policy (Movia, 2016). The first deployment was the improvements to the 5C line (which traverses Nørrebrogade) and includes selectively placed transit-only lanes and signal priority along the entirety of the route. Implemented in April 2017, this was Copenhagen's first transit priority project under these new approaches. Previously numbered 5A the route is one of the longest—from Copenhagen international airport in the east to Husum Torv at the municipal border to the west —and Denmark's busiest, carrying 20 million annual passengers (55,000 per day).

The goal of the scheme was to limit the mixing of buses with cars on Copenhagen's streets. This would reduce travel times and delay, and allow increased frequency to attract more riders. As bus fleets are routinely replaced with new vehicles, these new vehicles should have level boarding which speeds up loading and unloading and reduces dwell times at bus stops. Buses are also to have multiple doors and a larger floor area to ensure faster entry and exit and room for wheelchairs and prams. Bus stops include new windbreaks and digital traffic information. While bicycles are not allowed on board, some key suburban bus stops are outfitted with bicycle parking.

Movia's transit priority plan, like Copenhagen's cycle track expansion plans, has met with political conflict over the reallocation of car space, and especially opposition from the Right/Conservatives. In the 2017 municipal election campaigns, for example, the Conservative People's Party platform explicitly identified preservation of car space and "equal consideration between the forms of traffic" suggesting that cars were being maligned and car spaces unfairly removed. The Danish People's Party also expressed concern about finding a solution for the mixing of buses and cars that would not reduce car access.

For the Left/Progressives, the removal of car space is necessary, as exemplified by the successful outcome on Nørrebrogade. Once cars were restricted and bus-only lanes aligned next to widened cycle tracks, car traffic declined by 10 percent, cycling increased by 30 percent, and bus ridership also increased by 30 percent (Transport Advocate and Politician (1), 2016). When rebranded the 5C "City Line," the scheme reduced travel times by 17 percent along the entire route during peak hours and by 12 percent at other times (Movia, 2016). New articulated buses can carry 2,200 persons per hour along the entire route and it is anticipated that the line will experience a 5 percent increase in ridership (one million extra passengers annually).

Public transportation in Copenhagen is fragmented, with competing agencies with different values and goals. This shortchanges the public bus system. For example, Metro, a completely separate stand-alone agency with its own competing financial need for passengers, is expected to siphon off bus passengers when the Metro City Ring line opens in 2019. In 2015 Movia's buses transported 47 percent of public transit passengers in Copenhagen. Movia expects that figure to drop to 34 percent when the new Metro opens (Movia, 2016). If the bus system were left unchanged, the new Metro would take 34 million passengers annually from the bus system—just over 93,000 passengers per day.

In the longer-term Movia expects to lose 40 million passengers year-on-year in Copenhagen and the inner suburbs as more riders buy cars, shift to S-trains or to the Metro (and possibly cycle) (Movia, 2016). Some of Movia's highest ridership and most fiscally solvent bus routes will be impacted by the Metro City Ring. To stem these losses, and in addition to the transit priority proposals, Movia has proposed to restructure bus routes that loosely parallel or are close to the Metro City Ring line. This has raised concerns over social equity that resonate in the right to the city discourses that are important to the Left/Progressives.

Transit advocates fret that there will be unfair restructuring and service cuts to buses (Transport Advocate and Politician (2), 2016; Transport Advocate (2), 2016). Transit passengers will need to walk further to access public transit, and transfers between buses and Metro will lead to longer travel times, disproportionately impacting lower-income passengers, seniors, the disabled, and children. Some ideas floated by transit advocates include diversifying Movia's bus fleet with smaller, more nimble mini-buses to maintain services for the elderly, disabled and transit-dependent. These smaller buses would continue to operate on the bus lines experiencing large passenger shifts to the Metro. It remains to be seen until the Metro City Ring opens in 2019, but equity dimensions of public transit are a flashpoint that also indicates that Copenhagen is at a crossroads. If it remains an inclusive city, preserving and enhancing the bus system must have higher priority.

Disharmonies with the Metro

The interaction between the public bus system and the future Metro City Ring (see Figure 7.1) reflects a disharmony between two separate transit agencies. The Metro might also have considerable disharmony with cycling if the history of Metros in other European cities is an indicator and cycling trips switch to the Metro. More troubling from a green mobility perspective, is how much space above the Metro might be dedicated to cars once the new Metro City Ring opens in 2019?

Copenhagen's Metro is the newest transit system in the city. The Metro is a fixed-guideway electric train with short headways and stations that are further apart than bus stops, but closer together than commuter rail stations—making it competitive with both short cycling trips and bus trips. Operating on a third rail instead of overhead wires, with a uniquely short configuration of railcars, and automated as well, the Copenhagen Metro resembles an airport people-mover but with longer routes. Inaugurated in 2002, as of 2018 the Copenhagen Metro system was 21 kilometers long (or just 13 miles) with twenty-two stations, but the Metro transported 167,000 daily passengers in 2016 and has undergone a steady increase in ridership since its inauguration (Statistics Denmark, 2018).

Before the Metro City Ring line was approved some Left/Progressives in Copenhagen (and even a few fiscally prudent Conservatives in Frederiksberg) opposed the Metro and instead advocated for surface transit improvements for Movia, proposed light rail and the removal of car space (Transport Advocates, 2016). Invoking the famous transit priority system in Zurich (see Nash and Sylva,

2001 and Nash *et al.*, 2018), arguments made in favor of extensive surface transit modernization and expansion were that it would have been less expensive to build, it would have taken less time to build, it would have provided more geographically expansive and equitable service improvements, and would have been more dignified for passengers who would still remain above ground and could still interact with the city streets more readily.

Upgrading the existing bus system would have also improved transit for passengers in pre-existing corridors rather than creating new corridors. The underground Metro City Ring does not align with existing bus routes, for example. Left/Progressives opposing the Metro suggested that instead Copenhagen should have invested in converting trunk bus lines such as those on Nørrebrogade to electric tramway or light rail, and advocated for more transit-only lanes throughout the city. Fundamentally, this meant more car space would need to be reallocated, and therein lies the rub.

The decision to build a completely new transit system beneath Indre By and the Brokvarterene was made long ago and it is not the purpose here to rehash that debate. But it is important to remember that a decision to not modernize and implement transit priority on surface streets in Copenhagen, like the leftover spaces of historic Metros in other cities, was a political decision. There were several reasons behind this political decision. The financing of the Metro came from land sales, but upgrading Movia, city streets, or the S-train would have necessitated the redistribution of profits from land sales in the harbor to these modes instead of the Metro. Under the direction of Neoliberals and Right/Conservatives, the money from land sales could only be used for the Metro. The decision to couple land sales to the Metro was also consistent with the Neoliberal politics of mobility steering decisions about transit investment towards profitable land development instead for social services such as buses running to schools and hospitals, or to lower-income neighborhoods.

Labor politics was also a factor. An extensive above-ground transit priority network would need many operators and employees. The Metro was automated, and although it had station ambassadors and liaisons, it was expected to have fewer labor issues prevalent in public transit operations worldwide. For both the Neoliberals and the Right/Conservatives, most importantly, the Metro is underground or elevated above streets, and so therefore does not compete with car space. This last point must be further analyzed as the Metro City Ring approaches its opening in 2019.

In Chapter 3 we showed that cycling rates spiked in Copenhagen between 2012 and 2015 partly because of the construction of the Metro City Ring line (shown in Figure 3.1, referring to the chart in Chapter 3). The construction was a de facto car restraint mechanism that reduced driving in the city center. During construction there were many traffic choke points and car commuting declined from 33 percent in 2004 to 23 percent in 2014 (Oldenziel *et al.*, 2016). Cycling had become much more convenient and practical, and rates for trips to places of work and education by bicycle approached 45 percent (5 percent shy of the 50 percent goal). Considering the disruption to car traffic due to

Metro and utility construction, car restraint was proven effective and the city kept on functioning—significantly by bicycle.

However, as construction peaked and tunnels were finished, cycling declined to 41 percent—the experiment in car restraint was coming to a close. In anticipation of the opening of the new Metro City Ring line in 2019, there is speculation and worry by planners and advocates that the Metro might further thwart the cycling goals of Copenhagen by siphoning off or cannibalizing cycling trips in the urban core (City Transport Planner (1), 2016; Transport Consultant (2), 2016). Because the new Metro circulates entirely in the urban core it will almost definitely draw cyclists, but not car drivers, since there are already so few driving from point to point within the city (9 percent of all trips) (Transport Advocate and Politician (2), 2016). The Danish Road Directorate (2017) predicts that only 1 percent of commuting trips by car will shift to the Metro when the City Ring Line is completed, because most car trips in Copenhagen either begin or end outside Copenhagen, beyond the Metro service.

Some Copenhagen planners point out that there was a lack of cooperation and coordination between the Metro, an independent, stand-alone agency jointly owned by the Danish government and the municipalities of Copenhagen and Frederiksberg, and the city of Copenhagen's Technical and Environment Committee, which is responsible for planning for cycling (City Transport Planner (1), 2016). The Metro allows bicycles to be carried on board, but only during off-peak hours, and not for free, and the cost of a ticket is steep. It discourages the widespread carriage of bicycles on the Metro. The alternative, if passengers wish to combine cycling with the Metro, is to cycle to the station and leave the bicycle there. Copenhagen's planners are concerned that stations will have too little parking for cyclists.

Copenhageners have long managed without adequate cycle parking at train stations and as we suggested with the S-trains, parking is important but minor compared to larger political issues such as land use planning. More fundamental than station parking is, if, when the new Metro line opens and with construction no longer a hindrance, cars are let back into places that were previously off-limits to them, and Copenhagen makes it easier for motorists to drive (Transport Scholar (1), 2016; Transport Consultant (2), 2016).

The 2019 Metro City Ring opening is a key opportunity and inflection point for Copenhagen. Recall that today in the center of Copenhagen cyclists outnumber cars by about 10,000. If some cyclists shift to the Metro the balance on streets in the core might shift back to cars. Bus advocates also fret that planning after the Metro City Ring opens will mean that as bus routes are restructured and some routes removed, cars fill in the spaces (Transport Advocates, 2016). Motorists, having the slim majority on the streets, might demand more space, or at least resist further reallocations of space, a situation we have already detailed with parking, but that might spread.

Bicycle planners have asked: Once the construction is completed will streets that are currently closed to cars be reopened and will traffic increase near Metro stations? (City Transport Planner (1), 2016). Indications are that this will happen. Based on Copenhagen's 2016 Traffic Plan, which assumes that in the years ahead there would be the same levels of car traffic in Copenhagen that there were before construction of the Metro City Ring commenced. Copenhagen's traffic plans do not forecast less when the Metro City Ring opens (City Transport Planner (6), 2016). This means that all of those cars that disappeared will come back. A corollary flashpoint is how the spaces adjacent to the new Metro City Ring stations will be allocated (ibid.). Cycling to and from some Metro and S-train stations might become tense as cyclists navigate the car traffic around the stations.

Take as one example, Enghavevej in Vesterbro. Since construction of the Metro City Ring line began, Enghavevej has been jammed with cars despite also being an important north-south link for cyclists (City Transport Planner (6), 2016). Other parallel routes were closed during construction. Once the Metro is completed it is hoped that traffic will return to other routes, but there are no proposals to improve Enghavevej for cycling, despite the lack of cycle tracks at tough intersections such as Vesterborgade and Kingosgade.

What is likely is that cars will continue to use Enghavevej and also the routes that are reopened to cars—with a net increase in cars overall—because there has been little coordination between agencies about how to think about streets after the Metro opens. This may be a lost opportunity. Once the cars fill in, they gain the advantage of incumbency, making it politically more difficult to take away car spaces later. Spaces above ground might be too hastily returned to cars rather than reimagined for green modes (City Transport Planner (1), 2016; Transport Consultant (2), 2016).

There is also historic precedence of public transit dampening or cannibalizing cycling (Fietsberaad, 2010). As far back as the first bicycle boom of the 1890s and early 1900s, in some European cities there was deliberate effort to shift commuters from cycling to streetcars, and in the United States this occurred almost in totality (Friss, 2015). Cities like London and Paris built underground metros in the early twentieth century, and when transit was undergrounded, what was previously transit space was allocated to cars with little thought to cycling, and cycling declined (Carstensen and Ebert, 2012; Oldenziel et al., 2016). Decades later, Pucher et al. (2007) reported that Beijing saw declines in cycling beginning in the late 1990s due to a combination of subway expansion and residential relocation. The dispersal of residents from Beijing's Hutong, an optimal bicycle district like Copenhagen's Brokvarterene, to high-rise suburbs connected by the Beijing Metro, whittled down cycling rates, while Beijing also allocated street space to cars.

Yet it is important to emphasize that Metros did not kill cycling, politics killed cycling. In London, Paris, in most cities in the United States, and in Copenhagen as well, cycling can fit in very easily with public transit. The historic planning of metros in London or Paris could easily have been harmonized with cycling. Decisions to reallocate street space could have accommodated safe places to cycle, as these cities are realizing today. The decisions to allocate former transit spaces to cars was a political decision.

As Copenhagen's Metro City Ring opens in 2019 it will be a political decision to open streets to more cars, or keep streets more pedestrian and cycling oriented, with continued implementation of Movia's transit priority as well. Returning to the car-free center debate and harbor tunnel, spaces above new Metro City Ring could be coordinated with the creation of a more expansive car-free center, for example.

The Bispeengbuen removal debate

Reconsideration of the Bispeengbuen Motorvej (the Bispeeng Arch) must also be part of the broader car restraint debate. Bispeengbuen is a six-lane elevated viaduct that is part of a national road connecting Copenhagen's northwest suburbs to the city center. Carrying more than 50,000 cars and trucks daily, the Bispeengbuen bisects an area of dense apartment blocks in the Brokvartarene, straddling the northern boundary of Frederiksberg and Copenhagen, and following a curving pathway that was once a small creek called the Ladegårdsåen.

A short segment of Copenhagen's busiest crosstown artery, the Bispeenguen begins in a subsurface trench at Borups Allé, which connects further westward to the Hillerød motorway (see Figure 7.1). After passing beneath a set of S-train commuter rail tracks, the Bispeengbuen rises above ground and remains elevated for slightly less than 1 kilometer before touching back down at Agade. The crosstown road next transitions into the Åboulevarden, then H. C. Andersens Boulevard, and just beyond Copenhagen's City Hall the route crosses the harbor as the Langerbro, connecting to Amager Boulevard and finally Amagerbrogad.

The Bispeengbuen is a vestigial of the 1960s motorway era, and provoked controversy when it was built in the early 1970s. Elevating the roadway was thought to benefit adjacent neighborhoods because Frederiksberg and Nørrebro, beneath the viaduct, would still be connected by streets and pathways. Yet as with many similar elevated highways worldwide, the Bispeengbuen became an urban blight instead. The elevated roadway was very unpopular and it became a catalyst for rejecting other urban motorways in Copenhagen. In the early 2000s a noise shield, visible from the street level, was added at considerable expense, but the roadway was still noisy (Danish Town Planning Institute, 2004).

The Bispeengbuen is aging and will need to be substantially rebuilt in the coming years (Christiansen, 2018). In 2018 there was political discussion and emerging consensus to use the retrofit as an opportunity to replace the elevated roadway with an underground tunnel, thus reducing the blight of the highway while also daylighting the creek. In Copenhagen, the Red-Green Alliance, which oversaw the Technical and Environment Committee in 2018, promoted the tunneling concept, while in Frederiksberg, the Liberals also expressed support for tunneling. The Copenhagen consulting firms Cowi and Ramboll had each conducted separate explorations of tunneling as well.

Momentum increased in late 2018 as the Ministry of Transport, steered at the time by the Liberal Alliance, suggested that the government transfer the highway to the cities of Copenhagen and Frederiksberg and enable the municipalities to oversee tunneling. For the national government, transferring the Bispeengbuen offloads the full cost of rehabilitating to localities, which is in line with the neo-liberal hue of the Liberal Alliance (Christiansen, 2018).

To expedite the deal, the Transport Ministry has proposed contributing 125 million kroner (US $19 million in 2018) towards the cost of the tunneling. The two municipalities would match the Transport Ministry with local funds and assume responsibility for the project, perhaps as a municipal corporation like the Metro. Copenhagen's public utility, Hofor, may also contribute to the tunneling because it will involve stormwater management and stream restoration.

As the Bispeengbuen tunneling proposal gained traction and was vetted, the feud between Copenhagen and the national government over parking revenues came back to light. In October 2018 Copenhagen and Frederiksberg proposed that each municipality keep all of their parking revenue instead of sharing it with the Ministry of Economic Affairs and the Interior. This amounts to 200 million kroner (US $30.5 million) in on-street parking revenues kept in Copenhagen annually. Though yet to be solidified and awaiting a feasibility study (which was commissioned in late 2018) the parking revenue would then be dedicated to the Bispeengbuen tunnel project (Flensberg and Christiansen, 2018).

At first glance the replacement of the Bispeengbuen appears to be a popular idea with many benefits spanning the gamut of the urban livability discourse. Less noise and shadow from the elevated highway are the most obvious ones. Potential for green spaces in conjunction with new stormwater management is another. Yet the terms of the forthcoming agreement between the municipalities and the national government may not be so beneficial, depending on one's standpoint in the politics of mobility.

Significantly, if the proposals vetted in 2018 come to fruition, Copenhagen would be locked in to an expensive undertaking that favors cars in particular. The impetus for the project, the aging viaduct, means that something must be done however these issues are resolved. If the national government chose to simply rebuild the viaduct, the roadway would need to be closed and traffic rerouted over many months or years. It might be possible to retrofit in place and keep some lanes open, but at even more expense and even more time. Without doubt there would be great inconvenience and delay for cars on Copenhagen's busiest arterial.

If a tunnel could be constructed in advance of the closure, some of the disruption might be averted, benefiting car drivers. This makes tunneling attractive for the traditionally pro-car Neoliberals and Right/Conservatives such as the Liberals, the Liberal Alliance, and traditional Conservatives in Frederiksberg.

Yet the tunneling proposal requires that Copenhagen and Frederiksberg maintain the same capacity—six lanes of high-speed traffic—in the tunnel, meaning that the project is locked in for cars. If the terms are rigidly set, then narrowing the tunnel (as a kind of choke point to regulate traffic) is a lost opportunity, as is perhaps using some of the tunnel capacity for express bus transit. Enhedslisten has objected to the capacity lock-in, although other Left/Progressives have not found it to be a deal-breaker. The Socialist People's Party and the Social Liberals, for example, supported the deal as presented, maintaining six lanes, in early rounds of discussions (Flensberg and Christiansen, 2018).

The scheme to use parking revenues is also a lock-in for car-oriented infrastructure rather than redistributive to cycling or other green modes. While the tunnel would provide an opportunity for cycling or walking on the surface the long-term trajectory of maintaining a steady flow of cars into the city has proved difficult to manage as well as being detrimental to cycling and to Copenhagen's climate goals. Funds to the value of US $30 million annually would be a substantial boon, for example, to underwrite Plus-Net cycle track plans, many of which have been underfunded.

It also remains to be seen at what rate the national government might permit Copenhagen to raise the price of parking, which as we saw previously, is the source of a major political and ideological disagreement between the Neoliberals, the Right/ Conservatives, and the Left/Progressives. For example, the Red-Green Alliance expressed support for the parking component of the Bispeengbuen deal, but only if the municipality would then be free to also raise the parking rates as deemed fit by the city and without further interference by the national government.

Notably not part of the public discussions held in 2018, however, was the idea of simply removing the Bispeengbuen without replacing it. In the longer term, given the intractable problems of the car in the city, Copenhagen's shortcomings in meeting the city's climate and green mobility goals, and of the parallel vision of a car-free or car-lite urban core, one might ask why the complete removal of the Bispeengbuen would be accompanied by any roadway replacement at all? Should zero capacity be at least part of the feasibility studies and long-term planning for Copenhagen? Or as a compromise, less capacity and using the tunnel to meter or regulate the flow of cars in the city? As the Bispeengbuen tunneling proposal is studied in tandem with the harbor tunnel and Eastern Ring Road studies, would it not also be prudent to consider tolling the replacement tunnel, if built, instead of using parking revenue?

Conclusion: what kind of city?

Copenhagen's position as an iconic green mobility capital is at a unique and critical inflection point and as the debate over the harbor tunnel, the opening of the Metro City Ring line, and the Bispeengbuen demonstrate, there are many open-ended questions about what kind of city Copenhagen might become. Should the city build the harbor tunnel and Eastern Ring Road? Is new development attached to the tunnel and road the right place to build tens of thousands new houses? Is it wise to fill in new parts of the Øresund despite sea level rise, and how much will it cost to defend this new low-lying land from sea level rise? Would it not be more prudent to concentrate future housing inland on higher ground, perhaps by revisiting the Finger Plan and re-establishing regional cooperation and stricter land use planning?

Building the future Copenhagen inland also means deciding whether to build around the car or public transit. Will the region expand and improve the S-trains and other railways in a way that prioritizes bicycle geographies around the railway stations? Or, will the region expand the motorways and allow low-density, dispersed commercial and office development with abundant car parking?

Returning to congestion pricing and parking politics, there are even more questions. How will the city meet climate goals and maintain its green mobility status without the proven congestion charging approach? Will pricing simply offset reductions in the car tax, ultimately leading to more car purchases? Will the city and region save the bus system and improve it with transit priority streets? Will the city preserve its high rates of cycling or watch helplessly as it gets cannibalized by the Metro, and cyclists chased off the streets by more car congestion and pollution?

Obviously, the answers to these questions will depend on the ideologies and values that underpin the answer. Consider the first question asked here: Should Copenhagen build the harbor tunnel and Eastern Ring Road? The cost estimates for the full harbor tunnel and Eastern Ring Road, based on the 2013 Strategic Analysis Report, ranged from 19 billion to 22 billion kroner (US $2.9 to $3.3 billion in 2018) and escalated to as much as 30 billion kroner ($4.5 billion in 2018) depending on various construction scenarios and timelines. These cost estimates are likely much higher but as of 2018 no other estimates had been offered.

Taking US $27 billion kroner as a starting point for debate, in 2013 the local urban design outfit Copenhagenize presented a very different vision for the harbor tunnel and Eastern Ring Road. With the same amount of money (in 2013 currency), the city could complete the build-out of Copenhagen's Plus-Net bicycle system, build the entire regional cycle superhighway system, undertake a complete upgrade and automation of the regional S-train system, and build new tramlines to replace some of Copenhagen's busiest bus lines (Copenhagenize, 2013).

To be sure, the money forecast for the tunnel and bypass comes from a combination of toll revenue and land sales. Copenhagenize's outlays would need to be financed some other way. This points to other important questions about what kind of city Copenhagen might become. What would it look like if the above green mobility ideas from Copenhagenize were financed by a toll ring, increased parking pricing, and preserving the Danish car tax, while perhaps developing a transportation fund that opens land sales now restricted only to the Metro or the future harbor tunnel? What if portions of the existing harbor and Ørestad development schemes that have already been approved could be leveraged for green mobility, and likewise future development potential around S-train stations throughout the region also leveraged?

We have already seen a reluctance to charge a toll to enter the densest part of the city, and we see that the issue of parking pricing is politically volatile. The outcome of both the toll ring debate and Copenhagen's parking debate have been setbacks for green mobility and especially efforts to address bicycle capacity and Copenhagen's climate goals. The setbacks are not just a reversal on car restraint proposals, but they dampen the planning imagination. The intractable politics of the toll ring and parking diminish hope and lower expectations for a greener mobility future.

If the car tax is allowed to slide downward, this may also have a chilling effect on green mobility in Copenhagen. It remains to be seen if the Left/Progressives can dig in, and together with the Social Democrats re-establish support for the car tax. Some on the Left/Progressive side have proposed a car tax compromise. In exchange for either lowering or abolishing the tax, a carbon performance tax would be based on the emission of a car based on its full life cycle, from manufacturing, driving, to disposal (Transport Consultant (1), 2016). Yet details are unclear, and suffice it to say, caving in on the car tax all but guarantees that a carbon tax would collect less revenue.

Reconsidering Copenhagen's housing and economic development would need to also be part of these questions. With build-out of the housing unlikely to be realized until 2070, the newly rebranded tunnel-for-housing scheme at Lynetteholmen would lock-in the city for at least fifty years to a massive megaproject that contradicts green mobility visions. It leads to a need to rethink Danish laws regarding taxes. Currently the structure of taxes coerces municipalities to compete for residents because Denmark's municipalities impose a tax on income. Copenhagen loses revenue to suburban municipalities when the worker lives in the suburbs but sends commuters into the city. The city is compelled to seek raw land and build on landfill in the Øresund in order to maintain revenue streams.

What if Copenhagen and the region were to cooperate on housing issues, for example, and doubled-down on densifying the nodes around S-train stations, creating new bicycle cities along the railway lines? A scan of the region shows many large surface parking lots and other building opportunities that could accommodate just as many or more houses compared to what is being proposed in the tidal flats of the Øresund. Copenhagen and Denmark could go back to the Finger Plan and rework it, including revising the tax system so that municipalities are not attracted to fiscalized land use planning that misdirects future housing into poor locations.

Despite attempts by the Social Democrats to condition car restraint in the urban core to building the harbor tunnel, car restraint can be implemented without the tunnel, and also cheaply. The opening of the new Metro City Ring provides an opportunity for a much more expansive car-free or car-lite center that traces the ring itself. Other peer cities in Europe, such as Oslo, Norway, and Freiburg in Germany, are also implementing more expansive car-free zones based on green mobility politics.

Among transportation experts and advocates in Copenhagen ideas are germinating. Some have promoted ideas loosely based on Groningen in the Netherlands, which is a small university city with a city core car restraint policy that allows cars to enter and exit the core, but not to cross the core (Transport Consultant (2), 2016; Transport Scholar (1), 2016). Interestingly one of the earlier harbor tunnel reports examined the Groningen model, and the Ministry of Transport acknowledged that

Groningen arguably has one of the most successful and thorough car restraint policies in the world (Danish Ministry of Transport, 2013).

In Groningen, the city center is divided into four quadrants. If a car driver wants to cross the city center from one side to the other, they cannot. Instead, the car must be driven outward from the origin quadrant and the car must then circle back around the city center on an inner ring road just outside of the core. The car then re-enters the city center at the desired quadrant. This scheme effectively makes the center of Groningen nearly car free because it is simply not possible to drive across the center.

In Copenhagen, this kind of scheme could be implemented by using the lakes and harbor as a natural barrier to implement traffic calming so that cars cannot cross the center of Copenhagen, Indre By (Transport Scholar (1), 2016). Another more ambitious option would to extend the Groningen cordon model to the Ring 2 on the northwest, west, and southwest sides. The concept would make it such that it is not possible to drive from Østebro to Nørrebro or Frederiksberg without first going out to Ring 2, then back in (ibid.). Note that these boundaries roughly approximate to the city's preferred toll ring border during the debates over congestion pricing between 2008 and 2012 (see Figure 7.1). Inside the Ring 2 the scheme would include a 30 kilometer per hour speed limit on all streets (Transport Consultant (2), 2016).

For its part, some of Copenhagen's Left/Progressives, led by the Red-Green Alliance, have voiced support for more car restraint. Former Alliance Mayor Morton Kabell (2018) suggested that if the harbor tunnel and Eastern Ring Road were approved, the vague promises made by Social Democrats about car restraint were unacceptable. Instead, he suggested, in a grand bargain, that if the tunnel was being justified and designed to actually divert cars from the city center, the city must aggressively reallocate the streets in a more far-reaching way in the core. Radial roadways spanning from the center of Copenhagen outward should be reallocated to cycling and transit, and a greater proportion of Indre By made car free. This would be an opening to deploy the Groningen model as well.

A car-free or car-lite center, as with many other elements of green mobility, needs to be considered in conjunction with a broader right to the city framework. A very attractive city center which is car free in the spaces above the new Metro City Ring might become even more valuable than it is today. Already rents are high in Copenhagen, but from a neoliberal economic development perspective, making the center car free (with limited residential car access, of course) would likely lead to gentrification and displacement of low-income households. The right to a green mobility city will require intervention in the housing and land market, which is a tall ask, but not impossible if considering that Denmark has a history of strong social housing policies that could be reawakened.

References

Andersen, H. T. 2008. "The Emerging Danish Government Reform: Centralised Decentralisation." *Urban Research & Practice*, 1: 3–17.
Andersen, H. T. and Jørgensen, J. 1995. "Copenhagen." *Cities*, 12: 13–22.

Astrup, J. 2017. "Frank Jensen: Islamisk Stats tilbagegang gør terrorsikring nødvendig" (Frank Jensen: Islamic State's Decline Requires Terror Security). *Politiken*, 25 August.

Carstensen, T. A. and Ebert, A.-K. 2012. "Cycling Cultures in Northern Europe: From 'Golden Age' to 'Renaissance.'" In J. Parkin, ed. *Transport and Sustainability*, vol. 1: *Cycling and Sustainability*. Bradford, Emerald Group.

Cervero, R. 1998. *The Transit Metropolis: A Global Inquiry*. Washington, DC, Island Press.

Christiansen, F. 2018. "Minister Will Work for Money to Get the Bispeengbuen in Copenhagen into the Tunnel." *Politiken*, 25 October.

Colville-Andersen, M. 2012. *Outrageous Harbor Tunnel for Copenhagen*. Copenhagen, Copenhagenize Design Company. Available at www.copenhagenize.com/2012/11/outra geous-harbour-tunnel-for-copenhagen.html (accessed 30 June 2015).

Colville-Andersen, M. 2018. *Copenhagenize: The Definitive Guide to Global Bicycle Urbanism*. Washington, DC, Island Press.

Copenhagenize Design Company. 2013. *How to Spend 27 Billion Kroner*. Copenhagen, Copenhagenize Design Company. Available at www.copenhagenize.com/2013/02/how-to-spend-27-billion-kroner.html (accessed 24 November 2018).

Danish Ministry of Transport. 2013. *Østlig Ringvej Strategisk analyse af en havnetunnel i København* (Eastern Ring Road: Strategic Analysis of the Harbor Tunnel in Copenhagen). Copenhagen, Danish Road Directorate, Ministry of Transport.

Danish Ministry of Transport. 2018. *Lynetteholmen*. Copenhagen, Ministry of Transport. Available at www.trm.dk/da/ministeriet/lynetteholmen (accessed 10 December 2018).

Danish People's Party. 2018. *Traffic Policy*. Copenhagen, Danish People's Party. Available at https://lokal.danskfolkeparti.dk/Trafikpolitik (accessed 10 December 2018).

Danish Road Directorate. 2017. *Driving Forces: Why Is Road Traffic Growing in Denmark?* Copenhagen, Danish Road Directorate.

Danish Town Planning Institute. 2004. *Copenhagen General Plan 50 Years Later*. Copenhagen, Town Planning Institute.

Danish Transport and Construction Agency. 2016. *Public Transport in Denmark*. Copenhagen, Transport and Construction Agency.

Fietsberaad. 2010. *The Bicycle Capitals of the World: Amsterdam and Copenhagen*. Utrecht, Fietsberaad.

Flensberg, T. and Christiansen, F. 2018. "Parliament Lets Go of Hated 6-Lane Elevated Structure." *Politiken*, 15 November.

Friss, E. 2015. *The Cycling City: Bicycles and Urban America in the 1890s*. Chicago, University of Chicago Press.

Galland, D. and Enemark, S. 2012. *Danish National Spatial Planning Framework*. Aarlborg, Aalborg University.

Hagemeister, M. L. and Jensen, C. N. 2018. *Fantastisk dag for hovedstaden': Politikere roser nyt gigaprojekt i København* (Great Day for the Capital: Politicians Praise New Gigaproject in Copenhagen). Copenhagen, Danish Broadcasting Corporation, 5 October.

Homann-Jespersen, P., Sørensen, C. H., and Andersen, J. F. 2001. *Trafikpolitik og trafikplanlægning i Hovedstadsområdet* (Traffic Policy and Traffic Planning in the Capital Region). Roskilde, FLUX Center for Transport Research, Roskilde University.

Kabell, M. 2018. "Det er tid til at indfri løfterne, Frank Jensen, om en mere cykelvenlig by" (It's Time to Meet the Promises, Frank Jensen, to a More Bicycle-Friendly City). *Berlingske*, 10 November.

Kirkegaard, F. K. 2018. "Overborgmester: Trafikforhindringer i Københavns centrum skal lokke bilister i havnetunnel" (Lord Mayor: Traffic barriers in Copenhagen City Center will attract drivers to Harbor Tunnel). *Politiken Byrum*, 24 October.

Kraul, M. 2018. "Reaktioner på Københavns nye bydel: Flertal i Borgerrepræsentationen for Lynetteholmen" (Reactions in Copenhagen's new district: majority in the Borgerrepresentation for Lynetteholmen). *Politken Byrum*, 8 October.

London School of Economics and Political Science (LSE). 2014. *Copenhagen: Green Economy Leader*. London, LSE.

Mork, E. M. 2017. "Det kan godt være, at vi skal have en bilfri Indre By, men det skal ikke være for at terrorsikre" (It May Well Be That We Need a Car-Free Inner City, but It Should Not Be to Secure Us from Terror). *Politiken*, 25 August.

Movia. 2016. *Trafikplan 2016* (Traffic Plan 2016). Copenhagen, Movia.

Movia. 2018. *Statistics and Key Figures*. Copenhagen, Movia. Available at www.moviatrafik. dk/om-os/statistik-og-noegletal (accessed 15 June 2018).

Nash, A. and Sylvia, R. 2001. *Implementation of Zurich's Transit Priority Program*. San Jose, Mineta Transportation Institute, San Jose State University.

Nash, A., Corman, F., and Sauter-Servaes, T. 2018. *A Reassessment of Zurich's Public Transport Priority Program*. Paper Submitted to Transportation Research Board, 1 August,17.

Nielsen, T. A. S. and Skov-Petersen, H. 2018. "Bikeability: Urban Structures Supporting Cycling. Effects of Local, Urban and Regional Scale Urban Form Factors on Cycling From Home and Workplace Locations in Denmark." *Journal of Transport Geography*, 69: 36–44.

Norup, M. L. 2018. "Trafikforsker om ny ø i København: En god ide—men det kræver brugerbetaling" (Traffic Researcher on New Island in Copenhagen: A Good Idea—But It Requires User Payment). Copenhagen, Danish Broadcasting Corporation, 5 October.

Oldenziel, R., Emanuel, M., de la Bruheze, A. A., and Veraart, F., eds. 2016. *Cycling Cities: The European Experience*. Eindhoven, Foundation for the History of Technology.

Pineda, A. F. V. and Vogel, N. 2014. "Transitioning to a Low Carbon Society? The Case of Urban Transportation and Urban Form in Copenhagen since 1947." *Transfers: Interdisciplinary Review of Mobility Studies*, 4: 4–22.

Pucher, J. and Buehler, R. 2012. *City Cycling*. Cambridge, MA, MIT Press.

Pucher, J. and Kurth, S. 1995. "Making Transit Irresistible: Lessons from Europe." *Transportation Quarterly*, 49: 117–128.

Pucher, J., Peng, Z.-R., Mittal, N., Zhu, Y., and Korattyswaroopam, N. 2007. "Urban Transport Trends and Policies in China and India: Impacts of Rapid Economic Growth." *Transport Reviews*, 27: 379–410.

Risbol, S. 2017. "Ni københavnske spidskandidater diskuterede biler, boliger og bander" (Nine Copenhagen Candidates Discuss Cars, Homes and Gangs). *Politiken*, 15 November.

Statistics Denmark. 2018. *Metro and S-Train Increase Passengers*. Copenhagen, Statistics Denmark. Available at www.dst.dk/da/Statistik/nyt/NytHtml?cid=24612 (accessed 1 June 2018).

Skov, M. K. 2018. *COWI: Sustainable Urban Development Requires Traffic Solutions: Both Underground and Above Ground*. Copenhagen, COWI. Available at www.cowi.com/insights/tunnels-and-urban-development (accessed 7 December 2018).

Interviews held by Jason Henderson from 2015 to 2017 with transportation experts in Copenhagen

City Transport Planner (1), 2016.
City Transport Planner (6), 2016.
National Transport Planner, 2017.
Transport Advocate (2), 2016.

Transport Advocate (3), 2016.
Transport Advocate and Politician (1), 2016.
Transport Advocate and Politician (2), 2016.
Transport Advocates, 2016.
Transport Consultant (1), 2016.
Transport Consultant (2), 2016.
Transport Scholar (1), 2016.

CONCLUSION

Towards a politics of hope and the green mobility city

Politics of hope

Our goal in this book was to understand the politics behind making Copenhagen an iconic bicycle city because Copenhagen's cycling system is hopeful in a time of great discouragement.

Cycling, combined with compact cities, can replace many car trips, and have a transformative impact on our cities, our climate, our health, and our social systems. Copenhagen shows that cycling is practical, nimble, and thrifty. With 70 percent of households being car free and 62 percent of all trips to places of work and education made by bicycle, Copenhagen demonstrates that individuals and families with children can have a good life, cycle, and live car free.

Cycling has a spatial range and other characteristics that make it very suitable for replacing many short car trips in cities. When considering that half of all car trips in Europe are under 3 miles (4.8 kilometers) in length, and that in the United States 40 percent of all trips are under 3 miles in length, this could amount to a lot less driving and a substantial decrease in transport emissions. If harmonized with good public transit even more car trips can shift to cycling. Many cities in the world can achieve the cycling metrics of Copenhagen, and it is empowering to imagine what communities might look like if Copenhagen-levels of cycling were reached.

Copenhagen also shows that the shift from to cycling can be done rapidly. This is important because our window to meaningfully stay within the bounds of 2° Celsius or even 3° Celsius of warming is between 2019 and 2030. In that eleven-year climate window a fully connected and cohesive "Copenhagen style" bicycle system can be built in any city. Compared to cars and public transport, cycling infrastructure is relatively frugal, with far less demanding physical infrastructure and a shorter construction timetable.

Some observers, particularly those in the transportation justice advocacy community based in the United States, have cautioned about over-determining cycling infrastructure as a panacea and we agree (Lugo, 2018). There are deep inequities and social problems that cycling will not solve. Redistributive universal social welfare policies—something approaching the social democracy that undergirded cycling in Copenhagen—will be required.

This should trigger resolve in places like the United States and other global North countries to organize social democratic movements that might lay the groundwork for a green mobility transition. There were murmurs of this in the 2016 Democratic Party presidential primary, and again in the 2018 House of Representative elections in the United States. In early 2019 a "Green New Deal" program was introduced by some Democrats in the United States Congress (Friedman and Thrush, 2019). Invoking the 1930s New Deal, it combines government-financed public works, emphasis on equity, and climate mitigation. If a Green New Deal is to succeed it must include retrofitting US cities for cycling. In cities in developing countries in Africa, Asia, and Latin America, cities could focus on a redistributive, equitable, and socially inclusive politics of planning aiming for safe, secure, and just communities and cities. While the political tenants of social democracy might not apply seamlessly in all governance contexts, many of the principles can be adopted in the name of human wellness, safety, and security.

Emerging gazes towards an electric, driverless car future must not distract from the need to substantially reduce driving and the need to preserve and expand the opportunity for human-pedaled cycling. Electrification surely has a role in climate change mitigation, but assuming that the world (or Denmark) can simply switch from a petroleum-based mass car system to an electric (and driverless) mass car system is mistaken. A massive, emergency-scale, global wartime mobilization to renewable electricity is laudable (see Delina, 2016). Adding a huge outlay of electric cars would undercut that achievement.

Massive amounts of energy and resources would be claimed if the world shifted to a global electric car system, none of which has been scrupulously inventoried or adequately identified (Zehner, 2012). In Denmark, there are promises of huge new fleets of electric cars, namely by the Social Democrats, the Liberals, and the Liberal Alliance (Danish Ministry of Energy, Utilities, and Climate, 2018; Danish Social Democratic Party, 2018). Yet Copenhagen's 2025 Climate Plan notes that the Metro and S-trains will not operate fully on renewables energy until sometime in the 2030s, well after the 2019–2030 climate window closes. If the region cannot electrify green mobility based on renewables, how is it going to electrify cars?

Wind power for Copenhagen's climate goals, for example, have been contested (City Climate Planner (1), 2016). In 2018 Copenhagen had twenty-six wind turbines at its disposal through the public utility. The climate plan has a goal of at least 100 new wind turbines to produce hundreds of megawatts of electricity by 2020. None of them is near completion and most have yet to commence construction (Persson, 2018).

While renewables such as wind are popular, the reality in Denmark (and elsewhere) is that the proposals to scale-up renewables while perpetuating mass automobility are vague. Climate plans like Copenhagen's, considered one the world's best, could not address existing emissions from private cars. Instead, it offset car emissions by burning renewable biomass and then offset that stopgap measure with wind power that has yet to materialize. Furthermore, climate plans like Copenhagen's do not account for displaced emissions from electric car production and disposal—production of batteries and mining of rare earth minerals, not to mention the enormous amount of petroleum and coal that will fuel the transition. An attempted global switch to electric cars could exacerbate already potent inequities and would be extremely unfair to most of the world's population that would still not own cars.

Adding to the electrification conundrum are future driverless cars, which Denmark's Liberal Alliance especially anticipates (Danish Ministry of Transport, 2016). If driverless cars do become viable in urban traffic, and that is a qualified "if," these driverless cars will bring new claims on urban streets. During the period between the 2020s and 2050s, partially driverless cars might share parts of the roadways with conventional cars, while fully driverless cars may be limited to exclusive freeway lanes and automated valet parking garages. As of 2018 there was widespread and diverse speculation as to the rollout of fully automated, driverless cars (with no steering wheels or pedals). According to Schladover (2016) and McKinsey and Company (2016), the likelihood is a gradual rollout with significant overlap with conventional cars spanning decades.

Enthusiasts will clamor for driverless car lanes, noting that driverless cars may need to be segregated from conventional cars and "geo-fenced" within "driver-free" sectors of cities (Gagliordi 2016; Naughton, 2016). Uber, Google, and other driverless car magnates may pressure cities to cede space and resources: the right kind of roadways, with special markings, upgraded traffic signals, sensors, towers, poles, and antennae.

The political history of the car in countries with car industries suggests that space will not come at the expense of conventional cars, which will remain protected by a conviction that unfettered movement by car is a prerequisite of individual liberty and freedom. Rather, the most equitable and cleanest modes of transit—bicycles and public busses—may be forsaken.

Driverless cars may also escalate automobility and induce more metropolitan sprawl. Driverless cars might reduce household car ownership but concurrently increase the number of miles driven. Households owning two or more cars might shift to a single shared car, but that shared car would actually drive more as it shuttles back and forth, often empty, chauffeuring household members throughout the day (Schoettle and Sivak, 2015). The worst scenario is that driverless cars will be so easy to use that people will not mind longer trips or congestion because they can be entertained in the vehicle. Rather than reurbanizing around green transport, cities might physically expand more rapidly than ever before. Scenarios like these are especially disconcerting when considering the voracious appetite these new cars would have for future energy and resources.

Soaking up the planet's capital and resources to prop up the car system, especially given the 2019–2030 climate window, is the wrong path. There must be a political confrontation with the car—conventional or electric, human operated or driverless—and the policy options to mitigate climate change must include car restraint. To ignore this is folly and unjust, but here again Copenhagen provides some hope.

Based on the history of cycling, Copenhagen's bicycle system shows us what a city can look like without a disproportionately aggressive politics of car manufacturers. Copenhagen's cycling system can help to visualize what a city could look like if, through political change, the car industry was sidelined from positions of significant political influence. The absence of a car industry in Copenhagen should motivate doubling-down against the unequal political power that car industries (and now tech industries) have in countries like Germany and the United States and potentially increasingly in India, China, and Brazil.

In Copenhagen, when deindustrialization occurred during the 1970s and 1980s, there was also a solidly placed Left/Progressive political base in Copenhagen. It was positioned to imagine a bicycle city, and made it happen. At the same time, many cities in the United States, Canada, and Europe demolished city centers to make way for large roads and massive amounts of parking in response to deindustrialization. This should be instructive to advocates, planners, and environmentalists in cities around the world—be ready. Building a bicycle city, just as deciding to build a car city, is a political decision. When a city experiences economic or environmental calamity, building a cycling city is a very reasonable option for responding to crisis and rebuilding.

Defending hope

The municipality of Copenhagen deserves to be recognized as a leader in green mobility. Beyond Copenhagen and the rest of Denmark (perhaps except Odense and a few other important green bellwethers) that recognition is less, if at all, deserved. Copenhagen's cycling system and the ability to cycle safely for everyday life and to live car free is inspirational, but Copenhagen also shows that these conditions must be vigilantly guarded and cannot be taken for granted.

Some 90 percent of Copenhagen's transport emissions and pollution are from cars entering, leaving, or passing through the city (City of Copenhagen, 2015). "The city loves to cycle, the region likes to drive" (ibid., 2016) With no congestion pricing or other blunt instrument to reduce the inflow, and very heated parking politics, Copenhagen's climate plan is missing a keystone "stick" of car restraint, and (as of early 2019) the city still had no real ground plan for stemming the tide of more private cars into the city. Because of this omission, Copenhagen's world-class bicycle visions are sullied by capacity limits on cycle tracks. In 2018 there was a strong push to kickstart the car-centric harbor tunnel and Eastern Ring Road, and in 2019 the opening of the new Metro City Ring line might dampen cycling.

In all of this, the Social Democrat Party (both in Copenhagen and nationally) has had a puzzling and sometimes exasperating king-maker role when it comes to cycling and the car. On the most important flashpoints in Copenhagen's politics of mobility—the toll ring, car parking, the harbor tunnel—the Social Democratic Party has forsaken green mobility in favor of accommodating more cars in Copenhagen. The 2011 decision on congestion pricing hinged on the Social Democrats and they fumbled, betraying their promises to reduce car traffic, address climate change boldly, and make more space in the city for green mobility. This is confounding because core social democratic values such as egalitarianism, frugality, and social solidarity would ultimately limit the car.

It is not our intention to provide political calculus, but, consider if the Social Democrats did align with the Left/Progressives in Copenhagen. As the largest party in Copenhagen, but with just fifteen out of fifty-five seats in the City Council in 2018, the Social Democrats have no clear mandate to govern. The Social Democrats must choose allies to establish a governing coalition. The Social Democrats could align with the Red-Green Alliance and other left-wing parties, which hold twenty-two seats. If adding in the Social Liberals with five more reliable votes, a green mobility agenda could be readily implemented with a super majority of forty-two out of fifty-five City Hall seats.

Given that the Social Democrats have a solid left-wing heritage and have at times openly contested the car, it is not unimaginable to envision this. For Copenhagen this would be a pathway towards social democratic mobility. If Copenhagen's Social Democrats supported the Left/Progressive bloc on congestion pricing, Copenhagen would have a robust car management system as well as a progressive and redistributive source of revenue to share with the suburbs and to use towards building-out transit and regional cycling superhighways.

If the Social Democrats voted with the Left/Progressives on car parking policy, Copenhagen would probably have one of the world's most extensive parking management systems, and ample room to build the Plus-Net cycle track network and more transit priority lanes. If the Social Democrats opposed the proposed harbor tunnel and Eastern Ring Road, they could help to clear the deck on car-centric planning and redirect resources and political capital to expand public transit capacity on the S-trains and for the public bus system, Movia. Rather than rely on low-lying harbor landfill for a housing policy, the Social Democrats could spear-head renewed regional cooperation on housing. Instead, thousands of housing units could be built around rail transit nodes across the Copenhagen region, safely inland from sea level rise, and situated in bicycle geographies around pre-existing networks and routes like the S-trains.

However one looks at it, if the Social Democrats actually exercised social democratic politics, Copenhagen (and arguably Denmark) would be well on the way to expanding the cycling system and reinforcing the idea of future cities where cars make much less of an impact. Instead, the Social Democrats have elected to align with small Neoliberal and right-wing Conservative parties, promoting a car-centric mobility future for Copenhagen.

The Social Democrats might want to think further about the long game. Ultimately the goal of the Neoliberals—the Liberal Party and the Liberal Alliance—is to dismantle and privatize social welfare where possible, and this counters social democratic values. While the Social Democrats deploy neoliberal policies in the hope that these underwrite the social welfare system, dismantling social welfare is not a social democratic aspiration. Yet by bargaining on Neoliberal terms, the Social Democrats navigate precariously. The question is at what point do the Social Democrats back away from Neoliberal policies and shift back to their Left/Progressive roots? Even if the Social Democrats lost some voters to the Neoliberals or the the Right/Conservatives due to taking stronger positions on green mobility, the party could still piece together a strong governing majority with the Left/Progressives and the Social Liberals.

A core question for the Social Democrats, for both party representatives and for rank-and-file voters, is this: Is the car so important that it compels the party and its constituents to ignore all of the evidence of social, economic, and environmental problems attached to the car, while also sacrificing the social welfare system?

Ideology travels

The ambivalence of Copenhagen's Social Democrats reminds us that ideology travels, and that many of the same contradictions of Copenhagen's Social Democrats arise within similar political parties worldwide, but especially in North America and Europe. In the United States, many urban Democrats share this ambivalence, wavering from policies that promote bicycling and public transit to supporting middle-class car owners in ways that undercut cycling and public transit. In Democratically controlled cities around the nation, politicians defending free or cheap curbside parking block proposed cycle tracks, and oppose congestion pricing or higher gas taxes while simultaneously raising transit fares.

One of our aims in this study of Copenhagen was to also stimulate thinking and debate about how the politics of mobility travels between places and regions. By understanding the everyday politics of cycling in Copenhagen, other cities should be able to relate to the city. One can see that the politics of the car in Copenhagen is like the politics of the car in many cities. This should actually provide hope to city planners, environmentalists, climate experts, and urban right to the city advocates that this political struggle can have positive outcomes, needs to be constantly defended, and should not be taken for granted. By recognizing that the politics of mobility and political ideologies can travel, albeit variegated and nuanced, we can better situate and ground efforts to create a fair and just green mobility future worldwide.

Anyone concerned about the planet's climate future should not be dismissive of Copenhagen just because it is politically different and has a unique cycling system. Copenhagen is a special city to be sure, but the debates about the city's streets are remarkably similar to debates in cities around the globe. Copenhagen's politics of mobility is especially similar to debates in the United States and this should be instructive for the Danes.

The Liberals, the Conservative People's Party, the Danish People's Party, the Social Democrats, and the small but vocal Liberal Alliance—collectively these parties advocate for a more car-oriented society, and seek to impose this on Copenhagen. This would make Copenhagen more, not less, like the United States with high rates of car ownership and demands to make the city fit the car.

On the Left/Progressive side, the Red-Green Alliance explicitly advocates for car restraint and spaces in Copenhagen's urban core that are free of cars. The Alliance is frequently, but not always, joined by the Social Liberals, the Alternative, and the Socialist People's Party. They invoke a spirit of social democracy which is needed to confront climate change in cities.

Where there are dense urban cores, the places where cycling has the greatest potential, a strong dose of social democratic thinking about urban space, and specifically streets, may be effective in expanding green mobility. Consider for example, the vitriolic parking debates in just about every city in the United States, arguably in much of Europe, and increasingly worldwide. If public streets were conceptualized in a social democratic framework, cities would have more openings to rethink the public curb rather than simply handing it over as privatized car storage. In fact, anyone with a concern for climate change must especially rethink entitlement to park at the curb.

Given the discouraging politics on immigration and nationalist identity seen throughout Europe and the United States, it is also worthwhile to think what multicultural social democracy might look like, and how cycling fits in to that framework. In Copenhagen, there are many inclusive, multicultural dimensions of cycling, including courses available to many first- and second-generation Danes from countries without strong cycling cultures. Cycling courses in Copenhagen target immigrants and especially women who never learned to cycle (Red Cross Capital Region Copenhagen, 2018). A fully connected, cohesive, coordinated bicycle system, coupled to car restraint, can fit with many cities in the world and be equitable if also part of a redistributive and socially inclusive planning process.

Social democracy must also globalize to help to address the global inequity of the car. Private cars are disproportionately responsible for over a century of cumulative emissions and global discord and traffic violence, much of which originated in the global North. If the world meets the ambitious goals to stay within the bounds of 2° Celsius or even 3° Celsius of warming, while also mitigating enormous public health and economic inequities of the car system, German and American levels of car ownership cannot be replicated worldwide. Moreover, these two nations in particular should begin deep cuts in driving.

For cities in Africa, Asia, and South America, Copenhagen's cycling system is a model alternative to car-based planning and political promotion. Already dense African, Asian, and South American cities can fit the bicycle into the city quite easily if there is political will and citizen activism.

For green mobility advocates worldwide, it is important to incorporate right to the city thinking, but it is equally important that the right to the city movement prioritize cycling. Globally, urban-based left-wing movements frequently ignore mobility, or fail to connect the deep inequities of the car system, and thus undermine green mobility.

The world's left should look to Copenhagen's impressive Left/Progressive parties—the Red-Green Alliance, the Socialist People's Party, and the Alternative, as well as the Social Liberals—and take note. This is especially important for the intellectual and academic left, which has conducted a breadth of urban research that mostly ignores green mobility politics. It is time for the academic and climate left to come down to the street and get into the street fight, and imagine what social democratic mobility looks like.

Academics, climate scientists, urban planners, right to the city advocates, and elected officials all might begin to ask, discuss, and debate what does social democratic mobility look like in the twenty-first century, in a rapidly urbanizing world that is also rapidly warming? We have argued that looking back at Danish social democracy can start us out with some direction and inspiration, but by no means does it stop there. A research agenda is needed to map out what social democratic mobility looks like. As a final thought, we provide a few ideas.

Social democratic mobility includes the ability to live a comfortable, healthy, prosperous life car free. It means cities need to be designed to enable car-free living, and cycling, harmonized with public transport, is the key. Short daily trips must be made easy on foot or by cycling, and longer trips made easy by bus or train. The 1947 Finger Plan for Copenhagen can serve as a model of compact bicycle cities organized around the railways, and balanced with affordable and more flexible housing options. Affordable housing must also be conveniently located within these bicycle and transit geographies, so that car-free living is not just a boutique lifestyle for the select elite.

Car restraint is fundamental, and must be met either through outright bans and limitations or through redistributive congestion pricing. Systems that meter the flow of cars in order to reduce their volumes are ripe with technology. The Groningen model, expanded to larger cities, has great potential. Off-street parking in private residences, offices and commercial areas should be eliminated. Zoning, crafted to not just eliminate minimums but to create maximums that substantially reduce car space in cities, is also needed. At the curb, only minimal amounts of parking should be available; instead, curbs should be dedicated to green mobility.

Social democratic mobility will also have to say no to Uber and other privatized car-based systems with very dubious labor and business practices. Driverless cars, which may be technically feasible, should not be allowed to encroach on green mobility spaces, and any car whether driverless, electric, or human-powered fossil fuel, must be the very last option in rare circumstances such as remote areas.

Regardless of all this, if future generations write a history that documents human civilization's successful response to climate change, cycling will have had a significant role. Now is the time to act swiftly on climate change, and Copenhagen provides us with an important political script in terms of how to move forward.

References

City of Copenhagen. 2015. *Copenhagen Green Accounts 2014*. Copenhagen, Technical and Environmental Administration.

City of Copenhagen. 2016. *CPH 2025 Climate Plan: Roadmap 2017–2020*. Copenhagen, Technical and Environmental Administration.

Danish Ministry of Energy, Utilities and Climate. 2018. *Together for a Greener Future: Climate and Air*. Copenhagen, Ministry of Energy, Utilities and Climate.

Danish Ministry of Transport. 2016. *The Car Will Have a Greater Role in the Future*. Copenhagen, Ministry of Transport.

Danish Social Democratic Party. 2018. *Denmark Must Be a Green Giant Again: A Climate and Environmental Policy that Counts*. Copenhagen, Social Democratic Party.

Delina, L. 2016. *Strategies for Rapid Climate Mitigation: Wartime Mobilisation as a Model for Action*. New York, Routledge.

Friedman, L. and Thrush, G. 2019. "Liberal Democrats Formally Call for a 'Green New Deal,' Giving Substance to a Rallying Cry." *New York Times*, 7 February.

Gagliordi, N. 2016. "Ford CEO Promises Autonomous Vehicles for Mass Transit by 2021." ZDNet. San Francisco, CBS Interactive. Available at www.zdnet.com/article/ford-ceo-p romises-autonomous-vehicles-for-mass-transit-by-2021/ (accessed 15 November 2016).

Lugo, A. 2018. *Bicycle/Race: Transportation, Culture, Resistance*. Portland, OR, Microcosm Publishing.

McKinsey and Company. 2016. *Automotive Revolution: Perspective Towards 2030*. New York, McKinsey and Company.

Naughton, K. 2016. "Get Ready for Freeways that Ban Human Drivers." New York, Bloomberg Technology. Available at www.bloomberg.com/news/articles/2016-09-22/ robot-rides-may-force-error-prone-human-motorists-off-the-road?utm_content=busi ness&utm_campaign=socialflow-organic&utm_source=twitter&utm_medium=social& cmpid%3D=socialflow-twitter-business (accessed 15 November 2016).

Persson, S. 2018. "Copenhagen's Plan to Be CO2-neutral in 2025 Is Based on a Thin Thread." *Berlingske*, 17 June.

Red Cross Capital Region Copenhagen. 2018. *Cycle Training*. Copenhagen, Red Cross. Available at http://hovedstaden.drk.dk/det-goer-vi/integration/cykeltraening/ (accessed 2 August 2018).

Shladover, S. E. 2016. "The Truth about Self-Driving Cars." *Scientific American*, 314: 52–57.

Schoettle, B. and Sivak, M. 2015. *Potential Impact of Self-Driving Vehicles on Household Vehicle Ownership and Usage*. Ann Arbor, University of Michigan Transportation Research Institute.

Zehner, O. 2012. *Green Illusions: The Dirty Secret of Clean Energy and the Future of Environmentalism*. Lincoln, University of Nebraska Press.

Interview held by Jason Henderson in 2016 with transportation expert in Copenhagen

City Climate Planner (1), 2016.

INTERVIEWS IN COPENHAGEN

All the interviews took place in Copenhagen, in person, during summer 2015, fall 2016, or summer 2017. All interviews were conducted by Jason Henderson. Personal names and titles of interviewees were omitted.

City Climate Planner (1), 2016.
City Finance Executive, 2016.
City Transport Planner (1), 2016.
City Transport Planners (2 and 3), 2016.
City Transport Planner (4), 2016.
City Transport Planner (5), 2016.
City Transport Planner (6), 2016.
City Transport Executive (1), 2016.
National Transport Planner, 2017.
Regional Planner (1), 2016.
Transport Advocate (1), 2016.
Transport Advocate (2), 2016.
Transport Advocate (3), 2016.
Transport Advocates, 2016.
Transport Advocate and Politician (1), 2015.
Transport Advocate and Politician (1), 2016.
Transport Advocate and Politician (2), 2016.
Transport Consultant, 2015.
Transport Consultant (1), 2016.
Transport Consultant (2), 2016.
Transport Consultant (3), 2016.
Transport Scholar (1), 2016.
Transport Scholar (2), 2016.

INDEX

Note: Page numbers in **bold** refer to tables and in *italic* refer to figures.